D1235075

THE PHILOSOPHY OF
MATHEMATICS

THE PHILOSOPHY OF MATHEMATICS

AN INTRODUCTORY ESSAY

Stephan Körner

Emeritus Professor of Philosophy
University of Bristol and Yale University

DOVER PUBLICATIONS, INC.
NEW YORK

Published in Canada by General Publishing Company, Ltd., 30 Lesmill
Road, Don Mills, Toronto, Ontario.
Published in the United Kingdom by Constable and Company, Ltd.

This Dover edition, first published in 1986, is an unabridged and unaltered
republication of the second printing (1968) of the work first published in 1960
by Hutchinson & Co., Ltd., London.

Manufactured in the United States of America
Dover Publications, Inc., 31 East 2nd Street, Mineola, N.Y. 11501

Library of Congress Cataloging in Publication Data

Körner, Stephan, 1913-
    The philosophy of mathematics.
    Reprint. Originally published: London: Hutchinson, 1968. (Hutchinson
university library. Philosophy)
    Bibliography: p.
    Includes index.
    1. Mathematics—Philosophy.   I. Title.   II. Series: Hutchinson uni-
versity library. Philosophy.
QA9.K677   1986   510'.1   85-25438
ISBN 0-486-25048-2

To my wife

Just as everybody must strive to learn language and writing before he can use them freely for the expression of his thoughts, here too there is only one way to escape the weight of formulae. It is to acquire such power over the tool . . . that, unhampered by formal technique, one can turn to the true problems . . .

(Translated from Hermann Weyl's *Raum, Zeit, Materie*, 5th Edition. Berlin, 1923, §18)

CONTENTS

PREFACE

THIS essay is not intended as an introduction to Mathematical Logic or the Foundations of Mathematics, though mathematical inquiries and results are relevant to it. Where these are touched upon I have tried to explain them briefly and clearly, while avoiding technicalities as far as possible.

Of those topics which do fall within the field of the philosophy of mathematics, I have concentrated chiefly on (i) the relation between philosophical theses and the construction or reconstruction of mathematical theories and (ii) the relation between pure and applied mathematics. Chapters I, II, IV and VI are devoted to exposition of various views which are either historically important or widely held at the present time. Chapters III, V and VII are critical and lead up in chapter VIII to a proposal of a new philosophical position.

It gives me great pleasure to thank several of my friends and colleagues at the University of Bristol for their helpful comments and criticisms. My main obligation is to Mr J. C. Shepherdson, who has read the final draft of the whole book with great care and who has saved me from many imprecisions and at least one serious error. Professor H. Heilbronn has read the last chapter from the point of view of a pure mathematician and Dr D. Bohm from that of a theoretical physicist. These colleagues are not, of course, in any way responsible for my views or for any errors which remain for others to discover.

Professor J. W. Scott has done me the kindness of reading the typescript and of suggesting many stylistic improvements.

Finally, I should like to thank the editor of the series, Professor Paton, for the courtesy and understanding he has shown me during the writing of the book and, indeed, for asking me to write it.

INTRODUCTION

As the philosophy of law does not legislate, or the philosophy of science devise or test scientific hypotheses, so—we must realize from the outset—the philosophy of mathematics does not add to the number of mathematical theorems and theories. It is not mathematics. It is reflection upon mathematics, giving rise to its own particular questions and answers. Yet in spite of the distinction the connection between the two subjects is apt to be quite close. One cannot fruitfully reflect upon a subject if one has no acquaintance with it; and reflection on what one is doing may plainly be of profit by making the doing more efficient.

Throughout their history mathematics and philosophy have influenced each other. The apparent contrast between the indefinite flux of sense-impressions and the precise and timeless truths of mathematics has been among the earliest perplexities and problems not of the philosophy of mathematics only, but of philosophy in general; while philosophical accounts of mathematics in its relation to empirical science and logic have, on the other hand, suggested mathematical problems and even led to new branches of mathematics itself such as the non-Euclidean geometries and the abstract algebras of mathematical logic.

Since mathematical thinking is not only a highly specialized occupation but also part of the common business of living, the problems of the philosophy of mathematics will also be concerned both with what is generally familiar and with technical matters. This is not peculiar to the philosophy of mathematics. A similar division is found elsewhere in philosophy. Some of its problems, perhaps the most important of them, are with us habitually, and independently of any specialist training, whereas others arise only late, as it were, in a long and more or less arduous journey through some non-philosophical discipline.

Among the philosophico-mathematical questions familiar to everybody are some which arise from reflecting upon such statements as the following (the first three of which belong to pure mathematics and the remainder to applied):

(1) $1+1=2$;

(2) any (Euclidean) triangle which is equiangular is also equilateral;

(3) if an object belongs to a class of objects, say a, and if a is included in another class of objects, say b, then the object belongs to b.

(4) One apple and one apple make two apples;

(5) if the angles of a triangular piece of paper are equal then its sides are also equal;

(6) if this animal belongs to the class of cats and if this class is included in the class of vertebrates, then this animal belongs to the class of vertebrates.

Considering these statements, one will naturally ask such questions as: Why is it that they appear to be necessarily, self-evidently or indubitably true? Are they true in this peculiar way because they are asserted about objects of some special type—namely numbers, shapes, classes; or because they are asserted about objects in general or 'as such'; or are they possibly true in this special way because of their not being asserted of any objects at all? Is their truth due to the particular method by which they are reached or are verifiable—for example, an immediate and incorrigible act of intuition or of understanding? What is the relation between each of the three statements in pure mathematics and the one in applied mathematics which corresponds to it?

Reflection passes from more familiar mathematical matters to matters less familiar and more technical gradually and unavoidably. Any attempt to answer our questions about '$1+1=2$', for example, will force us to place this statement into the context of the system of natural numbers and, possibly, of still wider number systems. The questions we have asked about the apparently isolated statement will be seen immediately to extend to the system or systems to which it belongs. In a similar way we shall be forced to inquire into the pure system or systems of geometry and class-algebra, and into the structure of applied arithmetic, geometry and class-algebra. And this inquiry in turn will raise the question of the structure and function of pure and applied mathematical theories in general.

The full implications of a philosopher's answer to this last and central question will, of course, become clearer by considering the manner in which he deals with more specific problems, in particular controversial ones. One of these—one of the most important—concerns the proper analysis of the notion of infinity. The problem arises at an early stage of our reflection upon the apparently unlimited possibilities of continuing the sequence of natural numbers, and of sub-dividing the distance between two points; and it re-emerges at all later and more subtle stages of philosophizing about discrete and continuous quantities. If in the history of mathematics a new epoch can sometimes be marked by a new conception of infinite quantities and sets, then this is even more true of the history of the philosophy of mathematics.

We are now in a position to indicate in a preliminary way the topics of our present discussion. They are first the general structure and function of the propositions and theories belonging to pure mathematics, secondly the general structure and function of the propositions and theories belonging to applied mathematics, and thirdly questions about the role of the notion of infinity in the various systems in which it occurs.

The procedure to be adopted will be largely determined by the exigencies of an introduction. I shall start by sketching briefly the views which Plato, Aristotle, Leibniz and Kant held on our three topics. The reason for this is not to provide a historical outline—however incomplete. But these philosophers have frequently expressed in a precise and simple manner ideas which have formed guiding principles of modern schools of the philosophy of mathematics since the times of Boole and Frege and it seems natural to lead off from them.

The remaining chapters will be devoted to a critical examination of the logistic school, the roots of which go back at least to Leibniz; of the formalist school, some of whose main ideas are found in Plato and Kant; and of the intuitionist school, which also goes back to these two philosophers.

It seems desirable that the author of an introductory book should have views of his own on his subject; and that, if he has any, he should find room for some exposition of them. By so doing he will at least enable his readers to look in the right direction for possible misunderstandings and misinterpretations. I therefore conclude the book with some expressions of my own views.

Since the philosophy of mathematics is mainly concerned with the exhibition of the structure and function of mathematical theories, it would seem to be independent of any speculative or metaphysical assumptions. Yet it may be doubted whether such autonomy is even in principle possible—whether it is not even already restricted by the mere choice of a conceptual apparatus or terminology for dealing with the problems of the subject, or indeed by the type of problem considered important. In fact all the philosophies of mathematics so far put forward, and certainly those which will be discussed here, have been either explicitly developed within the framework of some wider philosophical system or have been pervaded by the spirit of some unformulated *Weltanschauung*.

Such general philosophical presuppositions show themselves most clearly when the exponent of a philosophy of mathematics is not content to draw attention to features which some mathematical theories actually possess, but maintains that all mathematical theories ought to possess them or—what amounts to the same thing—asserts that all 'good' or 'truly intelligible' theories do in fact possess them. The influence of general metaphysical convictions, in prescribing rather than describing, for example, the shape of a number system, comes out vividly in controversies regarding the admissibility or desirability, in such a system, of the notion of actually infinite aggregates, as opposed to aggregates merely potentially infinite. To confuse description and programme—to confuse 'is' with 'ought to be' or 'should be'—is just as harmful in the philosophy of mathematics as it is elsewhere.

I

SOME OLDER VIEWS

THERE is almost general agreement that, i n the second half of the nineteenth century, a new era in the philosophy of mathematics was initiated by the work of Boole, Frege, Peirce and some other mathematically minded philosophers and philosophically minded mathematicians. The period which begins with them has, for its most characteristic feature, recognition of the close relationship between the two fields, mathematics and logic; which two, strangely enough, had hitherto developed quite separately. The need for closer relations was sensed first by the mathematicians, in connection particularly with the theory of sets. The occurrence, there, of contradictions whose source was not clear, seemed to them to call for logical analysis; a task, however, to which the older logic was found unequal, being too narrow in its scope and insufficiently rigorous in its methods. New systems of logic free from these defects had to be developed. They embrace the types of deductive reasoning and of formal manipulation which are used in mathematics, and their precision is that of systems of abstract algebra. The new era is in fact dominated by attempts to clarify mathematics by means of logic, to clarify logic by means of mathematics, and to arrive thereby at an adequate conception of the relation between the two disciplines, if indeed they are two and not one.

The wealth of new ideas, new terminologies and symbolisms which accompanied the new ways of viewing mathematics and logic must not blind us to the elements of continuity between the pre-Fregean philosophy of mathematics and post-Fregean. The revolutionary changes have affected the tools of logical analysis to a greater extent than its purpose. It would be quite a mistake to say that the philosophical problems about the structure and function of systems of pure and applied mathematics, and the various fundamental attitudes towards these problems, have changed out of all recognition.

1. *Plato's account*

For Plato an important, perhaps man's most important, intellectual task was to distinguish appearance from reality. It is a task required not only of the contemplative philosopher or scientist but, even more, of the man of action, in particular the administrator or ruler, who has to find his bearings in the world of appearance and who must know what is the case, what can be done, and what ought to be done. To achieve order, theoretical or practical, in the world of appearances, which is always changing, we must know the reality, which never changes. Only in so far as we know that, can we understand and dominate the world of appearance around us.

The descent from this high and, so far, arid plane of philosophical generality to Plato's philosophy of pure and applied mathematics—and indeed also his philosophy of science and politics—presupposes that the distinction between appearance and reality can be made clear. In attempting this Plato follows certain suggestions arising from the Greeks' ordinary usage of the words corresponding to our 'appearance', 'reality' and their cognates. Plato did not thereby imply that ordinary usage may not also contain different suggestions or that ordinary usage is the ultimate standard of philosophical insight.

He noted that people habitually distinguish between a mere appearance and what is real without hesitation. Their judgments conform to certain more or less clear criteria. Thus we require of a real object that its existence should be more or less independent of our perceiving it and of the way in which we perceive it; that it should have a certain degree of permanence; that it should be capable of being described with a certain degree of precision; etc. All these requirements, in particular that of permanence, are susceptible of gradation and thus govern the use of the relative term 'more real than'. Plato is thus led to conceive of absolute reality and absolutely real entities as the ideal limits of their merely relative counterparts. The absolutely real entities—the Forms or Ideas—are conceived as being independent of perception, as being capable of absolutely precise definition and as being absolutely permanent, that is to say timeless or eternal.

G. C. Field, that most understanding and sympathetic of Platonic scholars, lays great emphasis on the naturalness of the transition from more-or-less concepts and criteria of reality to absolute ones.[1] This consideration not only offers a possible explanation of how Plato was led to the theory of Forms, but represents his central insight: the Forms include not only the ideal models of physical objects but the

[1] Field, *The Philosophy of Plato*, Oxford, 1949.

ideal states of affairs towards which it is man's duty to strive. In the present discussion, however, we are concerned only with the former and with these only in so far as they are related to Plato's philosophy of mathematics.

Which entities, we must ask, conform to the criteria of absolute reality? Not, certainly, the objects which make up the physical universe such as tables, plants, animals or human bodies. We could, however, think of some other interesting candidates for this high place, *e.g.* indivisible and indestructible bits of matter or of mind—*if* their existence could be demonstrated. If human souls are of that kind we might even hope for a proof of immortality. And we could think of some rather uninteresting candidates. We might, for example, consider any ordinary—more or less transient and indefinite—object, such as a table, and replace in our minds its transience by permanence, its indefiniteness by definiteness and its other 'imperfections' by the corresponding perfections. The result would then be the Form of the table; of which all physical tables are only imperfect copies. If this sort of Form strikes us as rather uninteresting it is, so I believe, because no reason can be adduced why there should not be a Form corresponding to every class of physical objects and every subclass of such a class until there is a Form for every single thing: not only a Form of a table, but also of a low table or a high one, of one covered or one bare, and so on.

There can be no doubt that at one time or another Plato regarded class-names such as 'table' as also being names of Forms. There are also, however, as Plato believed, some much more familiar entities than ideal chairs which conform to the rigid criteria to which an object must conform if it is to qualify as real or a Form: namely numbers and the objects of pure geometry, its points, lines, planes, triangles and the rest. Indeed a strong case can be made in support of the historical thesis that in the final stages of his development Plato admitted only two types of Forms: the mathematical and the moral.

Precision, timelessness and—in some sense—independence of their being apprehended is certainly, for Plato, characteristic of mathematical statements, and the view that numbers, geometric entities and the relations between them have an objective, or at least an intersubjective, existence is plausible. Speaking broadly we may say that Platonism is a natural philosophical inclination of mathematicians, in particular those who think of themselves as the discoverers of new truths rather than of new ways of putting old ones or as making explicit logical consequences that were already implicit.

Plato certainly held that there are mind-independent, definite, eternal objects which we call 'one', 'two', 'three', etc., the arithmetical Forms. He equally held that there are mind-independent definite eternal objects which we call 'point', 'line', 'circle', etc.—the geometrical Forms. In stating that one and one make two, or that the shortest distance between two points is a straight line, we are describing these Forms and their relations. Each of these has, of course, its multiplicity of instances; and some doubt and controversy has arisen about the status of such instances. The question has been raised as to what Plato's view is concerning, *e.g.* the double occurrence of 'two' in 'two and two make four' or the double occurrence of 'straight line' in 'two straight lines which do not have all points in common have at most one point in common'. Are the instances of 'two', *i.e.* the many twos with which the arithmetician operates, separate entities, and to be distinguished from the Form of twoness; or must one say that whatever is stated ostensibly about the many twos is ultimately stated about the unique Form? An exactly similar problem concerns the apparent instances of 'line'. According to Aristotle (by no means according to all later commentators), Plato did distinguish between (a) the arithmetical Forms and the geometrical on the one hand, and on the other (b) the so-called mathematicals, each of which is an instance of some unique Form—each Form having many such instances.

The question whether Aristotle misunderstood, or indeed whether he intentionally misrepresented his old teacher, will, I believe, be argued as long as there are Platonic and Aristotelian scholars. Without taking sides, it is worth noting that the relation between a mathematical concept, such as 'number' or 'point', and its apparent instances is by no means a trivial problem. It will confront us when we come to discuss the nature of mathematical existence-propositions.

There is, then, a world of Forms—timeless, mind-independent, definite objects—which is different from the world of sense-perception. It is apprehended not by the senses, but by reason. In so far as it contains the arithmetical and geometrical Forms it constitutes the subject-matter of mathematics. One of the strange features of mathematics, at least since Leibniz, is that, in spite of the certainty of its truths, it is by no means generally agreed of what, if anything, the true propositions mathematics are true. According to Plato they are clearly about something, namely the mathematical Forms. It is consequently quite easy to set down his answers to some of the questions which we listed in the introduction in our effort to demarcate, however roughly, the problems of the philosophy of mathematics.

The proposition that $1 + 1 = 2$, and all the other true propositions of arithmetic and of geometry, are *necessarily* true because they describe unchangeable relations between unchangeable objects, namely the arithmetical and geometrical Forms (or the equally unchangeable instances of these Forms). Their necessity is independent of their being apprehended by the discoverers of mathematical truths, independent of any formulation and thus of any rules governing a natural or artificial language. The truths of mathematics are equally independent of any preliminary act of construction. It is not essential, for example, to draw dots or lines on a black-board, or 'in one's mind', in order to be able to count or to perform arithmetical operations and demonstrations; and it is similarly not essential to draw triangles and squares in an empirical or non-empirical medium, in order to prove, say, the theorem of Pythagoras. Construction, according to Plato, is merely a practical need of the mathematician, or a guide he gives himself to discovery.

Plato's view about the relation between '1 + 1 = 2' and '1 apple and 1 apple make 2 apples', and in general about the relation between pure and applied mathematics, flows, like his account of pure mathematics, from his distinction between the reality of the Forms and the comparative unreality of the objects of sense-experience. These latter are only to some extent capable of precise definition or independent of the conditions in which we apprehend them (in perception). They are, moreover, not unchangeable although some of them do not change very much in certain respects during periods long enough to let us treat them as permanent. Thus if we compare the unchangeable, real object ONE with an apple, the latter can be properly said to be to a certain degree similar to, or, even better, to *approximate* to, the Form ONE. The technical phrases which Plato habitually uses are as a rule translated by saying that the apple—in so far as we apply arithmetic—*participates* in the Form ONE.

What has been said about the relation between one apple and the form of oneness applies similarly to the relation between, say, a round plate and the Form of circularity. We can treat the plate as if it were a geometrical circle because its shape approximates to the Form of circularity. This Form, like the Form of oneness, is apprehended, not by the senses, but by reason; that is, by grasping its mathematical definition or, as we might nowadays say, by understanding the equation of the circle.

For Plato, pure mathematics—it included in his day part of arithmetic and of Euclidean geometry—describes the mathematical Forms and the relationships between them. Applied mathematics describes empirical objects and their relations, in so far as they approximate to (participate in) the mathematical Forms and their relations. It might be tempting to say that the converse of approximation is idealization; and to regard the statement that some empirical objects and relations approximate to mathematical relations and objects, as being equivalent to the statement that mathematical objects and relations are idealizations of empirical ones. This, however, was not Plato's view. Plato considered mathematics not as an idealization, by the mathematicians, of certain aspects of the empirical world but as the description of a part of reality.

2. *Some views of Aristotle*

Aristotle's philosophy of mathematics is developed partly in opposition to that of Plato and partly independently of it. He rejects Plato's distinction between the world of the Forms, said to be true reality, and that of sense experience which is only to be understood as an approximation to the world of the Forms. According to Aristotle the form or essence of any empirical object, an apple or a plate, constitutes a part of it in the same way as does its matter. In stating that we see one apple or a round plate we do not, or should not properly, imply that the apple approximates in its empirical unity to a changeless and independently existing Form of oneness; or that the plate in its empirical roundness approximates to a Form of circularity.

Aristotle distinguishes sharply between the possibility of abstracting (literally 'taking away') unity, circularity and other mathematical characteristics from objects and the independent existence of these characteristics or their instances, *i.e.* units and circles. He frequently emphasizes that the possibility of abstraction does by no means entail the independent existence of that which is, or can be, abstracted. The subject matter of mathematics is those results of mathematical abstraction which Aristotle calls 'mathematical objects'.

About them, two uncontroversial statements at least can according to him be made: (a) each of them is in some sense *in* the things from which it is abstracted and (b) there is a multiplicity of them, *e.g.* there are as many arithmetical units, cases of two, three, etc., and as many circles, straight lines, etc., as are needed in calculation or geometrical argument. Other features of Aristotle's mathematical objects are, it seems, less clear: for example, the relation between one apple and mathematical unity, or a round plate and its mathematical circularity. Two possible interpretations of the relevant Aristotelian texts claim attention.

According to one main interpretation the empirical apple is one in

the sense that it is an instance of the mathematical universal 'unity',
just as it is red in the sense that it is an instance of the universal
'redness'. A variant of this interpretation would be to say that the
empirical apple is one, in the sense that it is a member of the class of
mathematical units, just as it is red in the sense that it is a member of
the class of red things. According to the other main interpretation the
empirical apple is one because it approximates to mathematical unity,
which we have *abstracted* from this and perhaps other objects. A
similar alternative presents itself if we wish to discuss the relation
between a round plate and geometrical circularity.

My own inclination is to accept the second interpretation. If we
accept it, then Aristotle's term 'taking away' would have to mean
idealizing abstraction or idealization, rather than simply abstraction.
His account of the subject-matter of mathematics would then be much
nearer to that of his teacher Plato than appeared at first sight. We
should have to say that whereas Plato holds mathematics to be about
Forms or, to use an equivalent term, Ideas existing independently of
the mathematicians, Aristotle holds it to be about idealizations performed by the mathematician.

Aristotle's view of the relation between pure and applied mathematics would then also come into clear light. The statements of applied
mathematics would approximate to statements of pure mathematics;
statements about drawn circles could be treated within a sufficiently
small margin, as statements about mathematical circles. Yet Aristotle
could not adopt Plato's theory that the reason why mathematical
statements are necessary is that they are descriptions of eternal and
independently-existent Forms.

Indeed it would not even be possible for him to speak of a true or
false idealization, but rather of one which would be more or less
adequate for some given purpose. Even if, however, a mathematical
theory is a set of idealizations, one need not be left without an account
of necessity. One might find the latter in the logical connection between
the various propositions of the theory. In other words, the necessity
would be found not in any single categorical statement about mathematical objects, but in hypothetical statements, statements to the
effect that if a certain proposition be true then a certain other proposition is necessarily also true. A leading authority on Greek mathematics, Sir Thomas Heath, who has gone with a fine tooth-comb
through Aristotle's works collecting all his statements about mathematics, confirms that for Aristotle the necessity of mathematics was
that of logically necessary hypothetical propositions. Evidence for this
view, as quoted by Heath, are a passage in the *Physics* and one in the

Metaphysics.[1] Aristotle's insight is even regarded as 'a sort of prophetic idea of some geometry based on other than Euclidean principles'.[2]

Aristotle also pays much more attention than Plato did to the structure of whole theories in mathematics as opposed to isolated propositions. Thus he distinguishes clearly between (i) the principles which are common to all sciences (or, as we might nowadays say, the principles of formal logic which are presupposed in the formulation and deductive development of any science); (ii) the special principles which are taken for granted by the mathematician engaged in the demonstration of theorems; (iii) the definitions, which do not assume that what is defined exists, *e.g.* Euclid's definition of a point as that which has no parts; and (iv) existential hypotheses, which assume that what has been defined exists—independently of our thought and perception. Existential hypotheses in this sense would seem not to be required for pure mathematics.

Aristotle's importance in the history of mathematical philosophy lies not only in his adaptation of the Platonic views to a metaphysics which does not need the reality of Forms and the comparative unreality of sensible objects. Neither does it lie only in his greater emphasis on the analysis of the structure of mathematical theories. Of more importance than either of these is the detailed formulation he has given of the problem of mathematical infinity; of which his discussion is still of great interest. Indeed he was the first of many to see the two main ways of analysing the notion of infinity as actual and as merely potential; and he was the first who made a clear decision in favour of the second alternative.

Aristotle discusses the notion of infinity in a connected passage of his *Physics*.[3] He distinguishes between the possibility of adding a further unit to the last member of any sequence of numbers, such as in particular the sequence of natural numbers 1, 2, 3, ... and the possibility of making always another subdivision of, say, a line between two points, which had previously been subdivided any given number of times. Here the possibility of going on *ad infinitum* is what may be meant by calling the sequence infinite, or the line 'infinitely' divisible (consisting of infinitely many parts). This is the notion of potential infinity. But one might also conceive both the notion of *all the elements of the sequence of natural numbers* and—which seems more difficult— *all the parts that are not further divisible of the line* as being in some

---

[1] *Physics*, II, 9, 200ᵃ, 15–19; *Metaphysics*, 1051ᵃ, 24–26.
[2] *Mathematics in Aristotle*, Oxford, 1949, p. 100.
[3] Book III.

sense given in their complete totality. This is the much stronger notion of actual infinity.

To try to present and analyse Aristotle's arguments for rejecting the notion of actual infinity would involve one in the finer points of history and Greek linguistic usage. Our interest is in the central idea behind the arguments; which seems to be that a method for a step-by-step procedure, *i.e.* for making the next step if the preceding step has been taken, does not imply that there is a last step, either in thought or in fact.

The rejection of the notion of actual infinity is regarded as being of little importance for the mathematician who, so Aristotle holds, needs only that of potential infinity for the purposes of mathematical demonstration. Whether Aristotle is right on this point is still a controversial issue. So also is the more radical view that the notion of actual infinity is not only not needed in mathematics, but is the unavoidable source of antinomies. This more radical thesis is less clearly expressed; and it might possibly be argued that Aristotle admits the possibility of the consistent use of actually infinite sets in a purely mathematical system which is not applicable to the physical universe.

3. *Leibniz's philosophy of mathematics*

Like Plato and Aristotle before him Leibniz developed a philosophy of mathematics because he was a philosopher in the wider sense of the term. He was the author of a metaphysical system of great beauty and profundity. He was also, indeed, a mathematician, a theoretical physicist, and many things besides. Moreover, all his intellectual activities and achievements were systematically interrelated. As it happened the system was never fully presented. In this respect he was more like Plato than like Aristotle. His great similarity to the latter lies in the close connection, one might even say the parallelism, between his logical and his metaphysical doctrines. Aristotle's position in logic, that every proposition is reducible to subject-predicate form, is paralleled by his metaphysical doctrine that the world consists of substances with attributes. Leibniz's more radical logical position, that the predicate of every proposition is 'contained in' the subject, is paralleled, on his side, by the famous metaphysical doctrine that the world consists of self-contained subjects—substances or monads which do not interact. The dispute amongst Leibnizian scholars as to which is the more fundamental, his logic or his metaphysics, testifies to the unity of his thought. Whatever may be said on either side, the view that he regarded either of the two as an unimportant appendix to the other seems highly implausible.

Unlike most modern philosophies of mathematics, that of Leibniz accepts the Aristotelian doctrine of the subject-predicate form of all propositions. None the less he anticipated modern movements, in particular modern logicism, by bringing logic and mathematics together. By a double innovation, he brought these, as yet quite separate, disciplines into relation. On the one hand he presents a philosophical thesis concerning the difference between truths of reason and truths of fact, and their mutually exclusive and jointly exhaustive character. On the other hand he introduces the methodological idea of using mechanical calculation in aid of deductive reasoning, not only within those disciplines which belong traditionally to mathematics, but also beyond them. This means, in particular, the introduction of calculation into logic.

Both for precision and brevity of the exposition one cannot do better than quote from the *Monadology* in which Leibniz, writing in 1714, two years before his death, gives a synopsis of his philosophy. 'There are', he says, 'also two kinds of *truths*, those of *reasoning* and those of *fact*. Truths of reasoning are necessary and their opposite is impossible: truths of fact are contingent and their opposite is possible. When a truth is necessary, its reason can be found by analysis, resolving it into more simple ideas and truths, until we come to those which have primacy . . . '[1] Truths of reason then are, as Leibniz puts it, grounded in the 'principle of contradiction', which he takes to cover the principles of identity and of the excluded middle. Not only trivial tautologies but all the axioms, postulates, definitions and theorems of mathematics, are truths of reason, *i.e.* they are '*identical propositions*, whose opposite involves an express contradiction'.[2]

Leibniz, as we have said, not only holds, with Aristotle, that every proposition is in the last analysis of the subject-predicate form; he also believes that the subject 'contains' the predicate. This must hold of all truths of reason which are of subject-predicate form and thus, according to him, it must hold of all truths of reason whatsoever. In what sense a truth of fact—say the truth that my fountain-pen is black—is to be regarded as having a subject containing its predicate is far less clear. Indeed, in order to explain the meaning of his assertion that the subject *of a truth of fact* contains its predicate, Leibniz has to bring in the notions of God and of infinity. The reduction of a contingent proposition, which will exhibit its predicate as contained in its subject, is only possible to God. Leibniz explains this by saying that, as in the case of surd ratios 'the reduction involves an infinite process and yet

1 Latta's edition, Oxford, 1898, p. 236.
2 *Op. cit.*, p. 237.

approaches a common measure so that a definite but unending series is obtained, so also contingent truths require an infinite analysis, which God alone can accomplish'.[1]

Another difficulty concerning truths of fact arises from the principle of sufficient reason 'which affirms that *nothing takes place* without sufficient reason, that is to say nothing happens without its being possible for one who should know things sufficiently, to give a reason which is sufficient to determine why things are so and not otherwise'.[2] This is for Leibniz not merely a general injunction to look for sufficient reasons to the best of one's ability, but in some ways, like the principle of contradiction, it is a principle of inference and analysis. The way in which it is to be applied, however, is not clearly determined; and in many cases, if not in all, only God actually does know things sufficiently to make its successful application possible.

It might seem that Leibniz's account of truths of fact, that is to say contingent propositions, is of no importance to his philosophy of mathematics. Philosophy of mathematics, however, is not concerned with pure mathematics only but also with applied. An account of applied mathematics must exhibit the relationship between mathematical and empirical propositions, and it might very well be affected by a mistaken or unclear view of the latter, or even by the lack of any view. This remark applies not merely to Leibniz, but to some of his modern successors.

Leibniz's conception of the subject-matter of pure mathematics is quite different from that of Plato or Aristotle. Mathematical propositions to him are like logical propositions in that they are not true of particular eternal objects or of idealized objects resulting from abstraction or indeed of any other kind of object. They are true because their denial would be logically impossible. Despite every *prima facie* appearance to the contrary, a mathematical proposition is as much or as little 'about' a particular object or class of objects as is the proposition 'If anything is a pen it is a pen' about my particular pen, or about the class of pens, or the class of physical objects or any other class of objects. We might say that both propositions are necessarily true of all possible objects, of all possible states of affairs, or, using the famous Leibnizian phrase, *in all possible worlds*. Any of these formulations is taken to imply the thesis that mathematical propositions are true and necessary because their negation would be logically impossible.

Whatever we may think of this, it is certainly clear—or at least as

---

[1] *De Scientia Universali seu Calculo Philosophico*, Latta, p. 62.
[2] *Principles of Nature and of Grace, founded on Reason*, Latta, p. 414.

clear as the notion of a proposition the negation of which is logically impossible or self-contradictory. We now turn to Leibniz's account of the relation between '1 + 1 = 2' and '1 apple and 1 apple make 2 apples' or, more generally, the relation between the corresponding propositions of pure mathematics and applied. We may of course evade the question by asserting that—as we wish to understand it— the second of the two propositions is (in some very strict sense) logically equivalent to the first, and that neither of them is either about apples or about the physical operation of putting them together, or indeed about the physical universe at all.

I do not propose to evade the question, and shall decide to understand the second proposition to be a proposition of applied mathematics or a very simple physics. This is because we know that such a decision is the one required in any consideration of the laws of Newtonian physics, of relativity, of quantum-theory and so on—unless we are prepared to argue that all mathematical physics or applied mathematics is *a priori* and contains no more information than does pure mathematics. Our decision has the advantage, in other words, of excluding non-philosophical difficulties from a question which philosophically is difficult enough.

Leibniz's philosophy of mathematics does not help much. According to it '1 + 1 = 2' (as a statement of pure mathematics) is true on the basis of the law of contradiction, and thus in all possible worlds; whereas '1 apple and 1 apple make 2 apples' (as a statement of physics) is true in this world which God was bound to create, if by the principle of sufficient reason, he was to have sufficient reason for creating it; *i.e.* if it was to be the best of all possible worlds. The relation between pure and applied mathematics is thus in a very direct manner—not merely in the 'last analysis'—given in theological terms. Leibniz's account of empirical propositions and consequently of the relation between pure and applied mathematics is no longer accepted even by those who in broad outline accept his view of pure mathematics.

If, in the eyes of those who accept his philosophy, Leibniz's analysis of the propositions of logic and mathematics brings these two subjects together, his methodological idea of introducing calculation into all subjects concerned with deductive relationships leads once again to a rapprochement between logic and mathematics—this time even apart from any special philosophical point of view. To Plato, as we have seen, the drawing of diagrams of various kinds and, as we may assume, the use of notational devices, were adventitious aids. They could be dispensed with. Leibniz on the other hand must have realized the practical impossibility of mastering really complicated deductions without a suitable symbolism. He must have found—in particular, in his researches into the possibilities of a mathematics of 'infinitesimals'—that the discovery of a symbolism for the representation of statements and demonstrations on the one hand, and the insight into their logical structure on the other, though separable in thought, are very rarely separable in fact.

The concrete representation, in suitable symbols, of a complicated deduction is, in his words, a 'thread of Ariadne' which leads the mind. Leibniz's programme is first of all to devise a method of so 'forming and arranging characters and signs, that they represent thoughts, that is to say that they are related to each other as are the corresponding thoughts'.[1] This idea anticipates exactly one of the central doctrines of Wittgenstein's *Tractatus Logico-Philosophicus*. It takes many forms in Leibniz's mind, one of which implies the arithmetization of logic, and reminds one of Gödel's famous method, to which we shall have to advert briefly in a later chapter.[2]

Once in possession of the *characteristica universalis* which represents thoughts in their relations to each other by symbols in corresponding relations, we need a method of symbolic reasoning or calculation; we need what is promised by—but does not quite follow—the title: *Calculus Ratiocinator, seu artificium facile et infallibiter ratiocinandi. Res hactenus ignorata.*[3] What Leibniz has to say about the symbolization of deductive reasoning is full of prophetic insights varying from the clear grasp of possible tasks to vague hints. A historian of philosophy, trying to demonstrate that there is nothing new under the sun, would be sure to find himself rewarded by reading through Leibniz's posthumous writings. But a critical introduction to the philosophy of mathematics such as we are undertaking must, where possible, deal with fully developed rather than seminal ideas.

4. *Kant: some of his views*

Kant's system of philosophy developed under the influence of the rationalist philosophy represented mainly by Leibniz and the empiricist philosophy represented mainly by Hume, and in conscious opposition to both. Both Hume and Leibniz divide all propositions into two exclusive and exhaustive classes, namely those of analytic and factual propositions; and both philosophers regard mathematical propositions

---

[1] Quoted from Becker, *Die Grundlagen der Mathematik*, Freiburg, 1954, p. 359.

[2] See *Elementa Characteristicae universalis*, Couturat's edition of Leibniz, Paris, 1930, pp. 42 ff.

[3] Couturat, *op. cit.*, p. 239.

as analytic.[1] However, Hume and Leibniz differ radically about factual propositions. About those of pure mathematics Hume on the whole says very little, and that not of great importance. Kant's philosophy of mathematics therefore, in so far as it is polemical at all, is directed mainly against Leibniz.

In order to go straight to the heart of it, and to show its connection with the rest of his philosophical system, it will be best to consider the threefold classification of propositions by which Kant replaces the Leibnizian and Humean dichotomy. His first class, analytic propositions (*i.e.* propositions the negations of which are self-contradictory), coincides with the analytic propositions of Hume and Leibniz. As regards non-analytic or synthetic propositions, he distinguishes between two kinds, namely those which are empirical or *a posteriori* and those which are non-empirical or *a priori*.

Synthetic *a posteriori* propositions are dependent on sense perception, in that any *a posteriori* proposition, if it be true, must either describe a possible sense perception (my pen is black) or logically mply propositions describing sense perceptions (all ravens are black). Synthetic *a priori* propositions on the other hand are not dependent on sense perception. They are necessary in the sense that if *any* proposition about the physical world, and in particular any proposition in physical science, is to be true, they too must be true. In other words, synthetic *a priori* propositions are necessary conditions of the possibility of objective experience.

This is not the place to give a critical discussion of Kant's argument in support of the thesis that some propositions are both synthetic and *a priori*. Neither can we consider his claim to have provided the premisses for a complete and systematic list of all such propositions, a list such as would remain unaffected by any changes in mathematics or the natural sciences. He divides synthetic *a priori* propositions into two classes: 'intuitive' and 'discursive'. The intuitive are primarily connected with the structure of perception and perceptual judgement, the discursive with the ordering function of general notions. An example of a discursive, synthetic *a priori* proposition is the principle of causality. All propositions of pure mathematics belong to the intuitive class of synthetic *a priori* propositions. We must, therefore, next turn our attention to them.

If we consider any perceptual judgement about the physical world, *e.g.* 'My pen is black', 'My pen lies between two pencils', it seems plausible to say that its truth or falsehood depends not only on definitions and the rules of formal logic, but on its correspondence or

---

[1] See Hume, *Treatise*, book I, part III, section XIV.

lack of correspondence with a perceptual situation which it describes. The relation between the concepts 'pen' and 'black' is not found by analysing 'pen' and 'black'; it is grounded in experience. It is equally plausible to distinguish with Kant, two aspects in every perception of, or proposition about, external objects, the empirical material which is located in space and time, and the space and time in which such material is located. If we assume that the structure of perceptual space and time remains unaffected by changes in the empirical material, and that there can be no perception unless located in time and no external perception unless located in both space and time, then we might take space and time to be the form of all perceptions and regard as the matter of perception all that does not belong to the form.

Being in space and time is a necessary condition of the possibility of perception, or at least, as Kant is wont to emphasize, of human perception. The question whether space and time are particulars or are general notions, in particular relations—whether they are more like physical objects or more like properties of, or relations between them —is answered by Kant in favour of the first alternative. His main reason is the difference between the sort of divisibility which belongs to particulars and the divisibility which belongs to general notions. To divide a particular, say an apple, is to cut it into pieces. To divide a general notion is to divide it into sub-notions. Space and time are divisible, so Kant holds, not as the property 'coloured' is divided into the various different colours, but rather as an apple is divided into pieces. Space is *rather like* a box and time *rather like* a stream.

Yet the space-box and the time-stream are particulars of a very special type. They are, as it were, the unchangeable containers *in which* the material of perception is found; they are not part of the changing empirical material of perception itself. In being unchangeable particulars, space and time remind one of Plato's Forms. But the analogy is not very close. Kant holds that they are not absolutely ('transcendentally') real. They are real only in so far as they are the conditions on which beings who are capable of perception and general thought can have objective experiences.

We can now see how synthetic *a priori* judgements of the intuitive kind are possible. For (a) in describing space and time we are describing particulars, which means we are making synthetic judgements; while (b) in describing space and time we are describing, not sense-impressions, but the permanent and unchanging matrices of them, which means that our descriptions are independent of sense-impressions, that is they are *a priori*.

Kant does not agree with the view of pure mathematics which

would make it a matter of definitions and of postulated entities falling under them. To him pure mathematics is not analytic; it is synthetic *a priori*, since it is about (describes) space and time. But if his account of mathematics had stopped here it could certainly not have hoped to explain the richness and variety of mathematics as then known. The description of space—perceptual space of course—could hardly have gone much beyond asserting it to be three-dimensional; or the description of time have gone much beyond asserting it to be one-dimensional and directed. Indeed, it would seem that Kant's influence on subsequent thinkers has been (largely) through his further development of the view that mathematical propositions are descriptions of space and time. To sketch briefly this further development, Kant will not allow that a full description of the structure of space and time requires mere passive contemplation. It presupposes the activity of construction. To 'construct a concept' is to go beyond proposing or recording its definition; it is to provide it with an *a priori* object. What Kant means by this is perhaps difficult, but by no means obscure or confused. It is quite clear what is and what is not entailed by constructing a concept. It does not mean *postulating* objects for it.

For example, the concept of a fifteen-dimensional sphere, which is quite self-consistent, cannot be constructed—even though we can (and must) *postulate* objects for it if we are to make the statement that in a 'space' of at least fifteen dimensions 'there exist' at least two such spheres with no 'point' in common. We can, however, construct and not merely postulate a three-dimensional sphere—or a circle (two-dimensional sphere)—in three-dimensional space. Its construction is made possible not merely by the self-consistency of the concept 'three-dimensional sphere', but by perceptual space being what it is. The *a priori* construction of a physical three-dimensional sphere must not be confused with the physical construction of, say, a sphere of wood or metal. Yet the possibility of the physical construction is based on the possibility of the *a priori* construction—the metal sphere on the possibility of a sphere in space—just as the impossibility of the physical construction of a fifteen-dimensional sphere is based on the impossibility of the corresponding *a priori* construction.

The distinction which Kant makes, in the introduction to the *Critique of Pure Reason*, second edition,[1] and elsewhere, between the thought of a mathematical concept, which requires merely internal consistency, and its construction, which requires that perceptual space should have a certain structure, is most important for the understanding of his philosophy. Kant does *not* deny the possibility of self-

[1] Ak. ed., vol. 3, p. 9.

consistent geometries other than the ordinary Euclidean; and in this respect he has not been refuted by the actual development of such geometries.

It is sometimes said that the use of a four-dimensional 'Euclidean' geometry in the special theory of relativity, or of non-Euclidean geometry in the general theory of relativity, has shown that Kant was wrong in holding that *perceptual space* is Euclidean. This is a more difficult point. I shall argue that he was indeed mistaken in assuming that perceptual space is described by three-dimensional Euclidean geometry. But I shall also argue that perceptual space is described neither by Euclidean nor by non-Euclidean geometry. The argument itself, however, must be postponed until other matters have been discussed.

Kant's account of the propositions of pure arithmetic is similar to his account of pure geometry. The proposition that by adding 2 units to 3 units we produce 5 units describes—synthetically and *a priori* —something constructed in time and space, namely the succession of units and their collection.[1] Again it is to be observed that the logical possibility of alternative arithmetics is not denied. What is asserted is that these systems would not be descriptions of perceptual space and time.

Kant's answers to our test questions about the nature of pure and applied mathematics can now be roughly formulated. The propositions of pure arithmetic and pure geometry are necessary propositions. Nevertheless they are synthetic *a priori* propositions, not analytic. They are synthetic because they are about the structure of space and time as revealed by what can be constructed in them. And they are *a priori* because space and time are invariant conditions of any perception of physical objects. The propositions of applied mathematics are *a posteriori* in so far as they are about the empirical material of perception and *a priori* in so far as they are about space and time. Pure mathematics has for its subject-matter the structure of space and time free from empirical material. Applied mathematics has for its subject matter the structure of space and time together with the material filling it.

Kant's notion of construction as providing the instances of mathematical concepts, the internal consistency of which is granted, assumed, or, at least, not called in question, has many recognizable descendants in later developments in the philosophy of mathematics. His analysis of infinity has been similarly influential. It reminds one in many ways of Aristotle's doctrine, except that in Kant the

[1] See, *e.g.*, *Prolegomena*, § 10, Ak. ed., vol. 4.

distinction between actual and potential infinity is worked out still more clearly.

In a mathematical sequence or progression a rule tells us how to take each step after the previous one has been taken. Kant will not allow the assumption that when such a rule is given the totality of all the steps is necessarily in some sense also given. The issue is particularly important in cases where there is no last step or where there is no first step. Consider, for example, the sequence of natural numbers, of which the first member is 0, and each further member is produced by adding 1 to its predecessor—it being presupposed that there are no other members in the sequence. The sequence as growing in accordance with the rule is quite different from the sequence as completed; and the statement that the process of producing further members of the sequence can be indefinitely continued does not entail that it can be completed or that the completed sequence can in this sense be regarded as given.

The Kantian distinction between potential infinity, or infinity *qua* 'becoming', and actual or complete infinity, is very similar to the Aristotelian distinction; but Kant's account of the notion of actual infinity differs considerably from Aristotle's. According to the latter not only are there no instances of actual infinity within sense-experience; it is logically impossible that there should be. Indeed Aristotle (like Aquinas later) attempts to demonstrate the existence of a First Cause by arguing that otherwise there would have to be an *actually* infinite sequence—which, he holds, would be logically absurd.

Kant does *not* regard the notion of actual infinity as logically impossible. It is what he calls an Idea of reason, that is to say an internally consistent notion which is, however, inapplicable to sense-experience since instances of it can be neither perceived nor constructed. Kant's view is that we can construct the number 2, and perceive 2 things; that we can construct the number $10^{10^{10^{10}}}$ even though we be incapable of perceiving so large a group of separate objects; and that, lastly, we can neither perceive nor construct an actually infinite aggregate.

The contrast between the actual infinite which cannot be constructed but is nevertheless 'needed', and the potential infinite which can be constructed (or exists in being constructed) is often emphasized by Kant. In the mathematical, and therefore constructive, estimation of magnitude 'understanding is as well served and as satisfied whether imagination selects for the unit a magnitude which one can take in at a glance, *e.g.* a foot or a perch, or else a German mile or even the earth diameter. ... In each case the logical estimation of magnitude advances *ad infinitum* with nothing to stop it.' 'The mind', however, Kant continues, 'hearkens now to the voice of reason which, for all given magnitudes, ... requires totality ... and does not exempt even the infinite ... from this requirement, but rather renders it inevitable for us to regard this infinite ... as completely given (*i.e.* given in its totality).'[1]

This transition from the notion of potential, constructive infinity to the notion of actual, non-constructive infinity is in Kant's view the main root of confusion in metaphysics. Whether it is required, desirable, objectionable or indifferent *within mathematics*, is a question which divides the contemporary schools of philosophy of mathematics perhaps more radically than any other problem.

[1] *Critique of Judgement*, § 26, Meredith's translation.

II

MATHEMATICS AS LOGIC: EXPOSITION

IN advocating calculation as a practically indispensable aid to all deductive reasoning, Leibniz expressed a methodological principle which has been adopted by modern logicians of every philosophical school. Fewer have endorsed his other thesis that logical and mathematical truths alike are grounded in the principle of contradiction and thus are all susceptible of being reduced to 'identical propositions' in a finite number of steps. Indeed, as it stands, this last position might be regarded as being not much more than a *credo*, and one which stands in need both of clarification if it is to become a practical programme, and of much hard and devoted work if it is to become a realized possibility.

Clarification is needed if we are to understand fully, either in what sense truths of reason are grounded in the principle of contradiction, or what the sort of 'reduction' is which would show them to be so. Leibniz himself seems to have regarded every truth of reason as being equivalent to a subject-predicate proposition of the form 'S is included in S or Q'; while, as regards the nature of the reduction, he seems to have assumed that it consisted in straightforward substitutions of the terms of the proposition *salva veritate* until the inclusion of the subject in the predicate was seen to take the self-evident form 'S is included in S or Q'.

The Leibnizian notion of an identical proposition—and, we may add, the similar Kantian notion of an analytic proposition—may seem *prima facie* somewhat too narrow if intended to cover the whole of logic and mathematics. Indeed, one might doubt whether the principle of contradiction itself is an identical proposition in Leibniz's sense. More seriously still, one might ask whether the principle of double negation, which Leibniz does regard as a truth of reason, is an identical proposition; for the logical validity of this principle (that p not only implies, but is implied by *not-not-p*) is, as we shall see in chapter VI, denied by some logicians.

These examples, in particular the second, disclose a certain obscurity not only in the notion of an identical proposition, but also in the notion of reduction concerned—reduction, that is, of a proposition not obviously identical to one which obviously is so. If there were no doubt about the nature of the reduction, one might at least attempt a settlement of, say, the dispute about the principle of double negation, by reducing it to an identity. But it is difficult to see how on Leibniz's showing one should set about such a task. Clarification is needed of the Leibnizian notions, both of an identical proposition and of reduction. Indeed there is room for an inquiry on the one hand into what he meant. and on the other into whether his notions can be replaced by similar ones, in terms of which the unity of logic and mathematics could be demonstrated.

1. *The programme*

Replacement was the path chosen by Frege, Russell and their successors. As a result, what in Leibniz was little more than a *credo* becomes in their hands a practicable programme. Frege, in particular, replaces the Leibnizian notion of an identical proposition—one in which the inclusion of the subject in the predicate is obvious, or can be made obvious in a finite number of steps—by his own notion of an *analytic* proposition: a proposition is *analytic* if it can be shown to follow merely from general laws of logic *plus* definitions formulated in accordance with them.[1] Similarly Frege replaces the Leibnizian reduction to identical propositions by his own procedure of *proving* an analytic proposition to be analytic. He does this by listing as clearly as possible not only all the fundamental logical laws which are permissible as premises but also all the methods of inference which it is legitimate to use.[2]

Frege's explanation of the analytic character of arithmetic presupposes that the general laws of logic, which he lists and uses as premises, are such as would generally and at once be recognized. These laws are propositions which he simply enumerates. He does not characterize them by any common feature, such as all analytic propositions might be presumed to possess though not always being immediately seen to possess it. A number of attempts have been made to provide a criterion of 'analyticity', especially in terms of the constituent parts of analytic propositions. An example of this is an

[1] See, *e.g.*, *Die Grundlagen der Arithmetik*, Breslau, 1884, § 4; also with English translation by J. L. Austin, Oxford, 1950.
[2] See, *e.g.*, preface to *Grundgesetze der Arithmetik*; also Frege-Translations by P. Geach and M. Black, Oxford, 1952, pp. 137 ff.

early attempt by Russell, later rejected by him as giving too wide a definition.[1] Similarly Frege replaces the Leibnizian reduction to identical propositions.

The path leading from the listed initial propositions, by inferential steps, to the theorems of arithmetic can be expected to be long, especially if every step is to be open to thorough inspection. For if an assumption is used even once which is neither one of the initial propositions nor a consequence of any of them, the demonstration is worthless. In order, therefore, to prevent surreptitious intrusion of non-logical assumptions, Frege and his followers adopted and extended the symbolic representation of deductive reasoning used by mathematicians. In this they were helped by earlier attempts to mathematize logical reasoning, in particular by Boole's treatment of the logic of classes.[2] The extension consists on the one hand of symbolizing not only the notions used in the traditional branches of mathematics but those used in all deductive reasoning; and on the other hand of formulating explicitly the permissible rules of inference. This means that every inferential step can be (a) represented by the transformation of one or more symbolic expressions into another and (b) justified by an appeal to clearly formulated rules.

It is clear that any demonstration to the effect that a particular theorem of arithmetic is analytic, *i.e.* that it can be deduced from the listed propositions of logic, involves changing symbols *en route*. The symbolic expressions marking the first stages will be obviously logical and will therefore contain only *logical* symbols such as those for propositional variables, negation signs, or signs indicating conjunction. (In symbolizing the principle of contradiction—that the conjunction of *any* proposition and its negation is false, obviously a logical principle—all these symbols are needed.) On the other hand the symbolic expressions marking the later stages, and certainly the last proposition of the formal deduction, will contain symbols which are not obviously logical, and are only seen to be logical as a result of the deduction. Somewhere in the path leading from the premises to, say, '$1+1=2$', the transition from obviously-logical symbols to symbols not obviously logical must occur.

And here an unavoidable problem arises concerning the nature and justification of the transition. Frege and Russell regard it as mediated by definitions. But their accounts of definition are different, and the difference is important for the philosophy of mathematics. According

---

[1] See *Principles of Mathematics*, London, 1903, chapter I, § 1, and, for the retraction, the introduction to the 2nd edition, London, 1937.
[2] See *The Mathematical Analysis of Logic*, Cambridge, 1847.

to Russell,[1] definition is a purely notational device. It is theoretically superfluous—a mere typographical convenience. 'A definition', he states, 'is a declaration that a certain newly introduced symbol or combination of symbols is to mean the same as a certain other combination of symbols of which the meaning is already known.' Yet, it is maintained that in at least two ways, definitions, theoretically superfluous though they are, often convey most important information. They imply 'that the *definiens* is worthy of careful consideration' and, further, that 'when what is defined is (as often occurs) something already familiar, such as cardinal or ordinal numbers, the definition contains an analysis of a common idea, and may therefore express a notable advance'.

Definitions, then, mere typographical (abbreviative) conveniences as they are, do not create new objects and do not ordinarily suggest the existence of such. Sometimes, however, a word which contributes to the meaning of certain contexts but has no meaning outside them does seem to refer to an object, *e.g.* the word 'Nobody' in the sentence 'Nobody runs as fast as I'. That this is not so is emphasized by the equivalence between the two statements 'Nobody runs as fast as I' and 'I am the fastest runner', and by the occurrence of the term 'Nobody' in one of these equivalent sentences and not in the other. There is no entity to which 'Nobody' refers, in the way in which 'Socrates' refers to a person. The point is that the definition of 'Nobody' is contextual; the term is defined by certain uses, or in certain contexts.

In speaking about individual numbers (as when we say the first prime number, the only number which . . .) or in speaking about classes of numbers (the class of integers divisible by two) we seem to be talking about something immaterial, logical, or mental. But if arithmetic is deducible from logic, then the deduced arithmetical propositions can hardly be assertions about objects of any kind; not, at any rate, if, as Russell holds, logic has no subject-matter. It (a) must be shown that the phrases which seem to represent entities (the so-and-so, the class of all things such that . . .), whenever they occur in the deduction of arithmetic from logic, are embedded in contexts, which do not imply the assumption that such mental objects exist; and (b) these contexts must be defined.

In his theory of descriptions Russell explains a method by which the phrase 'the so and so . . .', which seems to refer to an entity, can be absorbed into a context which need not refer to anything. The method is technically and schematically clear in his own example.

[1] See *Principia Mathematica*, 2nd edition, Cambridge, 1925, vol. 1, pp. 11 ff.

Consider the proposition 'the author of *Waverley* was Scotch'. This proposition is true only if the conjunction of the following three propositions is true: 'At least one person wrote *Waverley*; At most one person wrote *Waverley*; Whoever wrote *Waverley* was Scotch.' This method admits of various refinements and variations. It shows that the apparent application of a predicate to the author of *Waverley*, to the present King of France, the first prime-number, etc., can be explained without assuming that these definite descriptions describe any real entity at all.

Russell's way of dealing with the impression—correct or incorrect —that in talking about classes we are not talking about entities is similar to his theory of descriptions. Again, 'the class of objects which . . .', which seems to name an entity, is absorbed into contexts, each of which is defined as a whole and is free from the existential implications which Russell desires to avoid. The 'no-class' theory, as Russell's way of dealing with classes has been called, is less elegant than the theory of descriptions and less convincingly presented. It has come in for modification at the hands of later logicians who, however, still agree with Russell that mathematics is logic and that logic does not contain assertions about particular objects, whether physical, mental or logical.

On the function of definitions Frege differs greatly from Russell. His conception deserves attention not only because of its intrinsic interest, but also because the position that mathematics, though deducible from logic, yet contains assertions about (logical) objects, is one defended by certain contemporary logicians, in particular by A. Church.[1] The difference between the two branches of logicism, the nominalistic of Russell and the realistic of Frege, lies mainly in their different accounts of definition. If the difference is of little importance from the point of view of mathematical manipulation, it is—as both Frege and Russell insisted—of great philosophical importance.

According to Frege numbers are logical objects which it is the task of a philosophy of mathematics to point out clearly. To define them is not to create them, but to demarcate what exists in its own right. Contextual definition of logical objects will not do because it does not exhibit their character as independent entities.[2] To postulate them is according to Frege equally out of the question: we can postulate the existence of independent logical objects just as little as we can that of unicorns, which, if they existed, would exist independently of being

---

[1] See chapter I of his *Introduction to Mathematical Logic*, vol. 1, Princeton, 1956.

[2] See, *e.g.*, §§ 55, 56 of *Grundlagen der Arithmetik*.

postulated and which, as they do not exist, will not be brought forth
by any, even the most energetic, postulation.

Definition—in logic as in zoology—does not guarantee that the
notion defined will not be found empty. If the point of a definition is
to demarcate a class of objects, then these objects according to Frege
must be shown to exist. This is done by providing us with the means
of recognizing them. The principle which governs the identification
and recognition of logical objects is formulated by Frege as follows:
'I use the words "the function $\Phi(\xi)$ has the same *range of values* as the
function $\Psi(\xi)$" as having the same meaning as the words "the
functions $\Phi(\xi)$ and $\Psi(\xi)$ have the same value for the same argument." '[1]
Some of the terms in this formulation call for comment. A function
'$\Phi(\xi)$' is, as it were, unsaturated—the small Greek letter indicating
an open place to be filled by the name of an object. Frege calls '$\Phi(\xi)$'
a concept if the result of any such filling in yields an expression denoting a true or false proposition. The range of values of a concept contains all the objects, and only the objects, which fall under it. In other
words the range of values of a concept is its extension.

How the principle is used by Frege in the identification of logical
objects will be seen from the following examples. Let $\Phi(\xi)$ be the
function (concept) 'ξ is a straight line parallel to the straight line a'
and $\Psi(\xi)$ the function (concept) 'ξ is a straight line parallel to the
straight line b'. If now a certain straight line c belongs to the range
(extension) of both (falls under both) $\Phi(\xi)$ and $\Psi(\xi)$, then it has *something in common* with $\Phi(\xi)$ and $\Psi(\xi)$. The common characteristic which
has thus been discovered (pointed out) rather than postulated, may
now be defined. The *direction* of a straight line, say a, is the range of
values of the function 'ξ is a straight line parallel to a'.

Our second example is Frege's famous definition of 'number'. In
it the role played by the familiar notion of parallel lines is played by
the less familiar notion of 'similar' concepts. Two sets of objects are
similar if, and only if, a one-one correspondence can be established
between their members. The set of fingers on my hands is *in this sense*
similar to the set consisting of the toes on my feet, but not to the set
consisting of my eyes. (It is important to realize that the presence or
absence of the one-one correspondence can, it is supposed, be established without applying number-concepts.) Every concept determines,
as we have seen, a set of things, namely the set of things falling under
it—the extension of the concept. If the sets of things falling under two
concepts, *i.e.* their extensions, are similar we shall say that the concepts
themselves are similar, *e.g.* that the concept 'finger belonging to a

[1] *Grundgesetze*, vol. 1, § 3; Geach-Black translations, p. 154.

normal human being' is similar to the concept 'toe belonging to a normal human being'.

Now let $\Phi(\xi)$ be the function (concept) 'ξ is a concept similar to the concept a' and $\Psi(\xi)$ the function (concept) 'ξ is a concept similar to the concept b'. If now a certain concept, say c, falls both under $\Phi(\xi)$ and under $\Psi(\xi)$ then it has something in common with them: namely their number. Having demonstrated its existence—not postulated it— we may define the number of a concept, say a, as the range of values (extension) of the concept 'ξ is a concept similar to a'. It is as Frege insists most instructive to compare the manner in which one arrives at this definition of 'number' with that in which one arrives at the definition of 'direction'. In particular, this comparison will show that either both definitions are 'circular' or neither is. It will also show that the name given to the principle which is used to justify the definitions, the principle of abstraction, has been well chosen.

We shall see presently that Frege's principle of abstraction or some similar principle is needed for the execution of his programme. Yet its adoption among the obviously logical principles does not imply the adoption of the view that there are specifically logical objects. Indeed the method of contextual definition can be used—and has been used by Russell—to absorb the names of real or apparent entities arising from the application of the principle of abstraction into wider contexts in which they cease to appear as names of entities and become incomplete symbols, *i.e.* symbols defined in certain contexts only.

Although Frege and Russell differ in their conception of definitions and, consequently, in their conception of the principle of abstraction and in their views about abstract objects, their programme is in all other respects the same. It is, using Russell's words, to prove 'that all pure mathematics deals exclusively with concepts definable in terms of a very small number of fundamental logical concepts, and that all its propositions are deducible from a very small number of logical principles ... ' and to prove this with ' ... all the certainty and precision of which mathematical demonstrations are capable'.[1]

Some attempt must now be made to indicate how the programme above referred to has been executed. In undertaking this I shall endeavour to dwell on technical matters only in so far as their understanding is required for the elucidation of philosophical problems, especially those questions about the nature and function of mathematics, which are here our concern. It will be found important, throughout, to distinguish mathematical expertise from philosophical

[1] *Principles of Mathematics*, preface.

judgment, and avoid letting admiration for the former obscure possible defects in the latter.

The branches of logic—in the wide sense of the term needed by logicism—which have to be considered are briefly these: the logic of truth-functions, the extensional logic of classes, and the logic of quantification. The separation of these branches, while convenient and historically justified, is not necessary. Of all contemporary expositions of, and demonstrations within, the general programme of logicism, perhaps the most elegant are due to W. V. Quine.[1]

2. The logic of truth-functions

A true or false compound proposition, the components of which are also either true or false, is a truth-functional proposition (briefly, a truth-function) if, and only if, the truth or falsehood of the compound proposition depends only on (is a function of) the truth or falsehood of the components. Truth-functions are in a strict sense of the term, and not just metaphorically speaking, functions. To see this quite clearly let us consider as an example the familiar function $x + y$ or, written differently, $Sum (x, y)$; where—we assume—as the values of the arguments, we can take any natural number, and where the value of the function will again be such a number. Thus: $Sum (2, 3) = 5$. In a truth-function the values of the arguments are either *truth* (briefly, T) or else falsehood (briefly, F); and the values of the function, again, are either T or F. Thus, if p and q are propositional variables, then the conjunction (p and q) or $And (p, q)$ has the value T if, *and only if*, both p and q have the value T. We have: $And (T, T) = T$; $And (T, F) = F$; $And (F, T) = F$; $And (F, F) = F$. In other words, the conjunction of two propositions is defined as that truth-function of two propositions, which has the value T when both arguments have this value and which otherwise has the value F.

When the combination of two propositions is effected by means of the non-exclusive 'or'—the 'and/or' of legal documents—then it can be regarded as a truth-function of two arguments, and written $or (p, q)$. This function is defined by: $or (T, T) = T$; $or (T, F) = T$; $or (F, T) = T$; $or (F, F) = F$. We see that the value of this function is T if the value of at least one of its arguments is T; and F if the values of both arguments are F. The combination of two propositions by means of the *exclusive* 'OR'—*aut Caesar, aut nihil*—is defined as the truth-function: $OR (T, T) = F$; $OR (T, F) = T$; $OR (F, T) = T$; $OR (F, F) = F$. The combination of two propositions by means of '$\supset$'—often

---

[1] See, in particular, *Mathematical Logic*, revised edition, Cambridge, Mass., 1955.

misleadingly rendered as 'if . . . then'—is defined by: $\supset(T, T) = T$; $\supset(T, F) = F$; $\supset(F, T) = T$; $\supset(F, F) = T$. The third formula in particular, does not conform to many uses of 'if . . . then'. The combination of two propositions by means of ' $\equiv$ '—often rendered as 'if, and only if'—is defined by: $\equiv(T, T) = T$; $\equiv(T, F) = F$; $\equiv(F, T) = F$; $\equiv(F, F) = T$. Instead of *or* the symbol v, and instead of *And* the symbol &, are often used.

Other truth-functions of *two* propositional variables—making a total of sixteen—can be defined in the same way, although it is not always possible to find analogues in ordinary language. There is no reason why they ought not to occur there or, indeed, why one should not introduce them, if one so wished, into one's own ordinary language.

It has been shown that all truth-functions (strictly speaking all binary truth-functions, *i.e.* truth-functions of two arguments) can be introduced by definition if we start either (a) with the single notion of alternative denial, *i.e.* the truth function '$p|q$' which has the value F for '$T|T$' and the value T in the three other cases, or else (b) with the single notion of joint denial, *i.e.* the function '$p \text{ J } q$' which has the value T for '$F \text{ J } F$' and the value F in all other cases. (Note here that we have written the symbols for the truth-functional combination—'$|$' and 'J'—between the combined propositions. It makes no difference which order we adopt.)

The definitions of '*And* (p, q)', '*or* (p, q)', '*OR* (p, q)' may obviously be extended to similar functions of three or more arguments. Thus the truth-function of n arguments '*And* $(p_1, \ldots, p_n)$' has the value T only if all the arguments have this value; '*or* $(p_1, \ldots, p_n)$' has the value F only if all the arguments have this value, and '*OR* $(p_1, \ldots, p_n)$' has the value T, only if one and not more than one argument has this value. An obvious example of a truth-function of one argument is '*Not-*(p)', defined as having the value F if p has the value T, and the value T if p has the value F. (The tilde symbol $\sim$ is often used for *Not-*.)

It is important to see clearly which features are relevant to a compound proposition considered merely as a truth-function. For example, in so far as '(Brutus murdered Caesar) and (Rome lies in Italy)' is a truth-function, all that matters is the fact that both arguments of the function '*And* (p, q)' and therefore the function itself have the value T. *Qua* truth-function the compound proposition is *fully* represented by '*And* (T, T)', or more explicitly by '*And* $(T, T) = T$'. The proposition '(Lions are mammals) and (Elephants are larger than mice)', which is also one whose arguments are both true propositions, has exactly the same truth-functional structure as our first example. Indeed in so far as the two propositions are truth-functions they are both represented by one and the same formula '*And* (*T, T*)' or, more explicitly, '*And* (*T, T*) = *T*'.

Frege accounts for this situation—rather strange at first sight—by extending from concepts to propositions the distinction between connotation and denotation, or his own more precise distinction between sense and denotation (*Sinn* and *Bedeutung*). 'Rational animal' and 'featherless biped', to use a well-worn illustration, differ in *sense* but have the same *denotation*. In a similar way the proposition 'Brutus murdered Caesar' differs in sense from 'Lions are mammals', but the two have according to Frege the same denotation: they both denote *the true*, the truth-value *T*. In general every proposition denotes either *T* or else *F*, *F* being denoted by, *e.g.*, '2+2=5', 'Lions are fish' or any other false proposition.

Whether or not we accept this view—that all true propositions denote (are names of) the truth-value *T* and all false propositions denote (are names of) the truth-value *F*—is not the issue. The point is that in so far as we regard a proposition as a truth-function of its components, we must take account only of their truth or falsehood and ignore any other information which it or its components may convey.

Not every compound proposition can be regarded as a truth-function. Consider, for example, the compound proposition 'I believe that there will be no war in the next twenty years'. The truth or falsehood of this does not depend on the truth-value of the component 'There will be no war in the next twenty years'. Again the truth-value of '("all men are mortal and Socrates is a man" entails "Socrates is mortal")' does not depend on whether the antecedent and the consequent are true or false. The same applies to all assertions or denials of deducibility.

Lastly it is by no means obvious that every proposition is either true or else false. It is arguable that the English language contains indeterminate propositions; that propositions with indeterminate truth-value are desirable and available—*e.g.* for a satisfactory presentation of Quantum Mechanics[1]; that for deep philosophico-logical reasons the general validity of the law of excluded middle on which rests the dichotomy of propositions into those having the truth-value *T* and those having the truth-value *F* is unacceptable. Because of its importance for logic and the philosophy of logicism the point is worth

[1] See, *e.g.*, H. Reichenbach, *Philosophic Foundations of Quantum Mechanics*, Berkeley, 1948.

emphasizing that truth-functions are a very special and abstract kind of compound proposition, a type which in some respects represents *certain* features of English and other natural languages (see the examples above), but which in other respects are idealizations and simplifications (*e.g.* the assumption that there are two definite truth-values). As such they commend themselves for some purposes while not being suitable for others.

We can now turn to the problem which, from the point of view of logicism, is most important. Which truth-functions are logically necessary propositions, and thus can make permissible premisses for the deduction of arithmetic from logic? The answer to this is clear. A truth-function, say $f(p_1, p_2, \ldots, p_n)$ is logically necessary if, and only if, it is identically true, *i.e.* true for all values of the arguments $p_1, \ldots, p_n$. In other words, whatever the manner in which we fill the argument-places with T and F, the rule of composition, symbolized by f, is such that the truth-value of the compound proposition is always T. Consider for a simple example, the compound statement 'p or non-p' or, to take another familiar way of writing it, '$p \vee \sim p$'. Here if p is true, $\sim p$ is false and if p is false, $\sim p$ is true. One of the two component statements must be true. The case in which both are false cannot arise. Thus one of its member-statements must be true. Since an alternation in which one member is true must be true, *any* compound '$p \vee \sim p$', where p is *any* true-or-false statement, must be true.

Not all such identically true statements are so easily recognized as this. But so long as a compound statement is a truth-function consisting of a finite number of true-or-false components, it is possible in purely mechanical fashion to decide, after a finite number of steps, whether a compound truth-function is or is not true for all possible values of its arguments. (Methods of doing so are explained and illustrated in elementary text-books on symbolic logic.) The negation of a truth-function which is identically true is, of course, in turn, identically false, *i.e.* false for all values of its components.

Outside formal logic, truth-functions which are identically true or identically false are naturally of little interest. Thus there is seldom any point in saying, *e.g.*, that there either will be a war next year or there will be no war next year. In the ordinary business of conveying information we are concerned precisely with statements which are not identically true and are not identically false; with such statements, *e.g.*, as that there will be a war next year *and* nuclear weapons will not be used in it. This statement is neither necessarily true nor necessarily false.

Since the class of truth-functions which are identically true is well-defined and since it can be decided by routine methods whether any given truth-function is or is not identically true, there is no need to construct a deductive system the postulates and theorems of which would embrace all identically true truth-functions and no other propositions. Such a deductive system would be a so-called 'propositional calculus'. Many such systems have been constructed. Their chief value from our point of view is, as it were, pedagogical. They are simple examples, showing the rigour with which, in mathematical logic, theorems are deduced from postulates and definitions, according to rules of transformation. They form, moreover, the hard core of the more comprehensive and powerful logical systems within which we can deduce the theorems of arithmetic and other branches of pure mathematics.

What must be kept in mind at this stage of the discussion is that on the one hand *all* identically true truth-functional propositions (all truth-functional tautologies) are eligible as premisses in the attempted derivation of mathematics from logic; and that on the other hand we can show mechanically, in a finite number of steps, whether or not any given truth-functional proposition is a tautology. Within the domain of truth-functional logic the difference between obviously and not-obviously analytic (logically necessary) propositions is no longer important. For them the Leibnizian problem of *reducing* the latter to the former has been solved. All truth-functional tautologies are eligible premisses and some are needed as such. But we need other premisses in addition to them.

3. *On the logic of classes*

Among the propositions which by general agreement are regarded as belonging to logic, and therefore as permissible premisses in deducing pure mathematics from logic, are various principles concerning class-class and class-element relations. Their systematic presentation in the form of a deductive system, mainly due to Boole, preceded the similar presentation of the logic of truth-functions, and marks one of the sources from which modern logic sprang.

We shall first, following Lewis and Langford,[1] consider briefly the so-called elementary logic of classes, which will easily be seen to conform on the whole to our ordinary way of thinking about classes; but it has to be extended if it is to provide premisses sufficiently powerful to carry out the Frege-Russell programme. From out of the many systems of elementary class-logic, we choose one in which the following notions are accepted as being clear, namely: classes, symbolized by $a, b, c, \ldots$; the *product* of two classes, e.g. $a \cap b$,

[1] *Symbolic Logic*, New York, 1932.

which contains as members all elements common to a and b and no others; the universal class, symbolized by $\vee$, which consists of all the elements available in a given classified universe of discourse; the complementary class—*e.g.* a' of a which contains all those elements of $\vee$ which are not also elements of a.

In terms of these notions we define '$\wedge$' as '$\vee$'', *i.e.* as the empty or null-class; '$(a \cup b)$' as '$(a' \cap b')'$'; and '$a \subset b$' as '$a \cap b = a$'. '$(a \cup b)$' or the *sum* of a and b is the class which has as its elements every element of a, every element of b and every element common to both of them. $a \subset b$ expresses that every element of a is an element of b, *i.e.*, that a is included in b.

The following six formulae may then serve as postulates: (i) $a \cap a = a$, (ii) $a \cap b = b \cap a$, (iii) $a \cap (b \cap c) = (a \cap b) \cap c$, (iv) $a \cap \wedge = \wedge$, (v) If $a \cap b' = \wedge$, then $a \subset b$, (vi) If $a \subset b$ and $a \subset b'$, then $a = \wedge$. These formulae lead by straightforward reasoning to the theorems of elementary class-logic—just as in 'ordinary' algebra. Quite properly, text-books of formal logic substitute rules of transformation of formulae for the rules of informal reasoning.

To test the adequacy of the formalism to our intuitive reasonings concerning classes, we might interpret $a, b, c, \ldots$ as so many (sharply defined) classes of, say, animals, taking $\vee$ as the class of all animals and $\wedge$ as the empty class. Difficulties begin when we consider artificial classes (sets) whose elements, finite or infinite in number, are collected without reference to similarities or to the more familiar purposes of classification. Cantor, the founder of the general theory of classes or sets, defined a set (*Menge*) as 'a collection of definite, well-distinguished objects of our perception or thought—the elements of the set—into a whole'.[1] For him, it would be a set or class if, for example, we took the 'collection into a whole' of (a) all the irrational numbers (taken separately), (b) the *set* of all irrational numbers (taken as one element) and—for good measure—(c) Pythagoras.

One of the most important and fruitful events in the history of mathematical logic and the philosophy of mathematics was the discovery that Cantor's logic of classes, by admitting as a class any collection, however formed, leads to contradictions. Their occurrence, as we shall presently see, makes it necessary to distinguish between admissible and inadmissible classes, *i.e.* those which lead to internal inconsistencies and those which do not. The region of thought where such distinctions are imperative is a little like some notorious swamp which no one could drain, and which, consequently, it was

[1] 'Beiträge zur Begründung der transfiniten Mengenlehre I' in *Mathematische Annalen*, 1895.

imperative to bridge, by whatever artificial means were available. The path of deduction from logic to mathematics leads through this territory. Here it is where the followers of Leibniz, Frege and Russell are forced in order to cross from the one to the other, to make assumptions not 'obviously logical'—at least in the sense of 'logical' implied by Leibniz's, Frege's or Russell's use of the term.

We turn now to the well-known case of an antinomy in the theory of classes, discovered independently by Russell and Zermelo.[1] If we admit, as sets, all collections of objects which conform to Cantor's definition, *i.e.* all collections whatever, we must *ipso facto* admit sets which contain themselves as elements. For example, the set consisting of all sets is itself a set, and thus an element of the set of all sets—itself; or again the set of all sets with more than ten elements has more than ten elements and thus again contains itself as a member. These sets may strike us as being somehow abnormal as compared with normal sets or classes which do not contain themselves as elements: for example the set of all men—which contains every man but not also itself as an element.

Let n be the set of all *normal* sets, *i.e.* the set of all sets not containing themselves as elements. If m is any particular normal set, say the set of all men, then m is an element of n—or in symbols $m \in n$. Again it would be correct to state of any abnormal set a (*e.g.* the set of all sets) that $\sim(a \in n)$.

If $m \in n$ (if m is a normal set) then, and only then, $\sim(m \in m)$ (m is not an element of itself). We now ask whether n—the set of all normal sets—is normal. If $n \in n$ (if n is a normal set) then, and only then, $\sim(n \in n)$ (n is not an element of itself). If $\sim(n \in n)$ then, and only then, $(n \in n)$. In other words the statement that n is normal implies, and is implied by, the statement that n is not normal. The antinomy can be deduced in the ordinary logic of classes if we define within it the class n of all normal sets and its complementary class $n' = a$ of all abnormal sets.

What is needed for the deduction of mathematics from logic, is a logic of classes which can provide principles suitable for this purpose without their being also liable to lead to antinomies. Also needed is a justification of these principles on better than merely pragmatic grounds. The programme of logicism is to deduce mathematics from logical principles, not from principles some of which are logical and some not. Unless the premisses can be shown to be logical

1 See Fraenkel, *Einleitung in die Mengenlehre*, reprinted by Dover Publications, New York, 1946, or Fraenkel and Bar-Hillel, *Foundations of Set Theory*, Amsterdam, 1958.

principles, the work is not done. There can of course be nothing but admiration for a deductive technique which should deduce the whole of arithmetic and more from a very small set of postulates by formalized rules of inference. But unless the said small set consists of postulates either obviously or demonstrably logical, it does not demonstrate the truth of logicism.

Russell argues that the antinomy we have been discussing, and some others which can be constructed within the framework of Cantor's theory of sets, result from a certain kind of vicious circle. The principle which shows what is to be avoided and how it can be avoided is formulated in *Principia Mathematica* as follows.[1] ' "Whatever involves *all* of a collection must not be one of the collection"; or conversely, "If, provided a certain collection had a total, it would have members only definable in terms of that total, then the said collection has no total." ' The principle suggests a hierarchy of types, and restrictions on the formation of classes in terms of it. We can distinguish classes from one another according to type as follows: type 0, individuals; type 1, classes of individuals; type 2, classes of classes of individuals, etc. Apart from these classes of pure type, classes of mixed type are also conceivable.

If we define a normal set as one which does not contain itself as an element, there is always one way of getting round the troublesome '$n \in n$, if and only if $\sim (n \in n)$'—(n being the set of all normal sets). We may simply prescribe that a class must never contain itself as a member. The prescription might be strengthened by prescribing that any class must contain only classes of lower type as members; or that any class must contain only such members as are of the immediately lower type. This last rule, that if a class is of the nth type its members must all be of the $(n-1)$st type, has in fact been adopted by Russell.

Although the vicious-circle principle avoids the antinomies it leads to difficulties which require new postulates for their obviation. The vicious-circle principle stratifies classes into types, and the stratification extends to all propositional functions—the propositional function $\phi(x)$ being the class of all objects of which $\phi(x)$ can be asserted with truth; thus it extends to all propositions, since to assert a proposition is to apply a propositional function, or (to use Frege's expression) a concept, to an object. (Strictly speaking we should of course also have considered the function: $\phi(x_1, x_2, \ldots, x_n)$ where $n > 1$.)

Russell is well aware of this. 'It is important', he says in *Principia*

[1] 2nd edition, vol. 1, p. 37.

Mathematica[1] to observe 'that since there are various types of propositions and functions ... all phrases referring to "all propositions" or "all functions" or to "some (undetermined) proposition" or to "some (undetermined) function" are *prima facie* meaningless ...' He emphasizes this point. 'If mathematics is to be possible', he says, 'it is absolutely necessary ... that we should find some method of making statements which will usually be equivalent to what we have in mind when we (inaccurately) speak of "all properties of *x*".' We need not here consider the method proposed by Russell, or the further postulates in which it is embodied. Nor need we say anything about the various efforts of Russell's successors to reduce the epicycles, so to speak, which Russell found must be imposed on the vicious-circle principle, itself an epicycle on Cantor's theory of sets.

4. *On the logic of quantification*

The last group of postulates needed to carry out the Frege-Russell programme concern the use of the terms 'all' and 'some' in mathematics. These were the last part of the logical apparatus to be formalized. The need for them becomes very clear when we consider the transition from statements in which a property is asserted of one individual object, to statements in which the same property is asserted of a finite number of objects, and thence to statements in which it is asserted of an actually infinite number.

Consider the statement 'Socrates is mortal'. Here we may distinguish with Frege (see above, p. 37) (i) the unsaturated function 'x is mortal' and (ii) 'Socrates' as the name of one of its values. Since the saturation of the function by the name of an object leads to a true-or-false proposition, the unsaturated function is called by Russell a 'propositional function' and by Frege a 'concept'. Assuming, as we are free to do, that the number of men is finite and that all of them are unambiguously distinguished by names, then to assert 'all men are mortal' is from our point of view the same as to assert 'Socrates is mortal and Plato is mortal and ...', where the dots suggest the completion of a very long conjunction of propositions which yet are finite in number and are all derived from the same propositional function in a simple, straightforward way. This conjunction is a truth-functional combination of elements which has the value *T only if* all the components have the value *T*. Otherwise it has the value *F*. We may abbreviate the very long conjunction by writing: '(x) (x is mortal)' or more schematically '$(x)f(x)$'. (For convenience we assume that our universe of discourse consists only of men.)

[1] Vol. 1, p. 166.

To assert that there exists, say, a man, is to assert 'Socrates is a man or Plato is a man or . . .' where the dots suggest the completion of a very long alternation, derived from the same propositional function in an obvious manner. This alternation is again a truth-function which has the value F only if all components have the value F. We may write it in the form '$(\exists x)$ (x is a man)' or more schematically '$(\exists x)\ \phi(x)$'. In other words our universal and existential propositions can perfectly easily be incorporated into the logic of truth-functions, so long as the ranges of our propositional functions are finite.

Now some of the most important propositional functions in pure mathematics, such as 'x is an integer' or 'x is an irrational number', are of ranges which are not finite. They must be regarded as infinite, at least potentially. The philosophical mathematicians who adopt the Russellian programme regard the ranges of both 'x is an integer' and 'x is an irrational number' as actually infinite, and they regard the latter range as being in a clearly definable sense greater than the former. If, therefore, rules for the use of 'for all x' and 'there is an x such that . . .' are to be formulated, they cannot be regarded as rules for such truth-functional combinations as involve the connectives 'and' and 'or' (in symbols '&' and 'v'). In point of fact truth-functional conjunctions and alternations have been used nonetheless as heuristic analogies to universal and existential statements.

Thus in a universe of discourse consisting of two objects, say a and b, such a proposition as '$f(a)$ and $f(b)$' can be written '$(x)f(x)$'. Now, since '$(f(a)\ \&\ f(b)) \supset f(a)$' is a truth-functional tautology (see p. 39) the formula '$((x)f(x)) \supset f(a)$' is in our finite universe of two objects also a truth-functional tautology. If now the range of '$f(x)$' is infinite we may still regard '$((x)f(x)) \supset f(a)$' as valid, by making it a postulate of our logical system or by seeing to it that it becomes deducible as a theorem. In a similar way we have for our finite universe of two objects a truth-functional tautology '$f(a) \supset (f(a)\ \text{v}\ f(b))$'. Since a and b are the only objects, this formula can be written as '$f(a) \supset ((\exists x)\ f(x))$'. If our universe is infinite we must, if we want to be able to assert the formula in general, include it among our postulates or theorems. The heuristic introduction of the principles of quantification by extending them from finite to 'infinite' ranges of propositional functions[1] is most instructive in showing how very strong our 'logic' must be for mathematics to be deducible from it.

Postulates providing the infinite ranges covered by the logicist equivalents of the phrases 'for every integer . . .' or 'for every real

[1] Carried out in detail by Hilbert and Bernays, *Grundlagen der Mathematik*, Berlin, 1934 and 1939, vol. I, pp. 99 ff.

number . . .' are susceptible of various interpretations. They may be regarded as merely technical devices, admissible so long as they are demonstrably innocent of leading to contradictions. This is essentially the view of Hilbert and his school. They may be regarded, again, as inadmissible because misrepresenting the nature of mathematics. This is essentially the view of Brouwer and his disciples. They may be regarded, lastly, as empirical assumptions about the world. Thus Russell regards as an empirical hypothesis the statement that there is an infinite class of individuals in the universe. To say that there are fewer than nine individuals in the world would be to assert a false empirical proposition, to say that there are more than nine would be to assert a true empirical proposition. To say that there is an infinite number of individuals in the world would according to Russell be to make an empirical statement which may be true or false but which in *Principia Mathematica* is assumed to be true.[1]

5. *On logicist systems*

Any logicist system has to be judged both from the mathematical and from the philosophical point of view. Mathematically, we must ask whether its symbolism is as precise and its deductions are as rigorous as can, in view of existing mathematical techniques, reasonably be demanded, or whether the system indeed represents an advance on these. Philosophically, we must confront and compare the logicist system with the logicist philosophical theses and programmes, as enunciated by Leibniz, Frege, Russell and others. We must judge the system in the light of the thesis that mathematics is logic (in various senses of this aphorism) and also judge how far this thesis is illuminated by the system or, perhaps, obscured by it. If the system is defective *qua mathematics* its confrontation with philosophical theses and programmes may sometimes be pointless. Yet the mere mathematical perfection of the system is not sufficient to validate a logicist philosophy of mathematics.

Since the present essay is concerned with mathematical logic only in so far as the latter is relevant to the philosophy of mathematics, I shall not attempt to criticize any logicist—or other—system from a mathematical point of view. I shall always assume, or grant for the sake of argument, that the formal system discussed is mathematically sound, or can without substantial alteration be made so. To accept, say, Russell's mathematics while rejecting his philosophical thesis that mathematics is deducible from or translatable into logic is to do

[1] See also *Introduction to Mathematical Philosophy*, 2nd edition, London, 1920, pp. 131 ff.

nothing more out of the way than, *e.g.*, to accept Euclid's mathematics while disputing the philosophical thesis that perceptual space is Euclidean.

Every logicist system draws its list of postulates and rules of inference from the logic of truth-functions, the extended logic of classes and the logic of quantification. The list of postulates and the list of inference rules are not independent. For example, a suitably large list of postulates enables one to economize in inference rules. Sometimes an infinite number of postulates can be adopted, *e.g.* by stipulating that all truth-functional tautologies, or all formulae conforming to certain schematic descriptions are to be postulates. This way of specifying the postulates is adopted in Quine's *Mathematical Logic*. *M.L.*—as this system is often called—needs on the other hand only one rule of inference, namely the *modus ponens*: If ($\phi \supset \psi$) and ϕ are theorems, then ψ is also a theorem.

M.L. is one of the most influential systems constructed in the light of the logicist ideals. Its aim is to improve on *Principia Mathematica* by avoiding some of its difficulties, especially those connected with the theory of types. Since our interest lies in logicism as a philosophy of mathematics and since we wish to grant the mathematical claims of *M.L.* or similar systems, as far as possible, it is well to state what Quine himself claims for *M.L.*

He claims that the notions of arithmetic can be defined in purely logical terms; that 'the notions of identity, relation, number, function, sum, product, power, limit, derivative, etc. are all definable in terms of our three notational devices: membership, joint denial, and quantification with its variables'.[1] Definition here may be both explicit or contextual, and does not imply the existence, in any sense, of objects falling under the defined concepts.

He does not claim to have deduced the theorems of arithmetic from purely logical principles. *M.L.*, like all systems which are intended to incorporate (substantially) the whole of classical mathematics, contains among its postulates a principle limiting the free use of phrases such as 'All *classes* such that . . .' or 'There exists a *class* such that . . .', since this freedom leads to contradictions. In *M.L.* the freedom of quantification with respect to classes, classes of classes, etc. is restricted by a rule for the stratification of the universe of discourse which is simpler than Russell's theory of types. It is not claimed to be a logical principle.

This is what Quine has to say about the various ways of achieving the logicist aim and of avoiding contradiction: 'The least artificial and

[1] *M.L.*, p. 126.

at the same time the technically most convenient formulation would seem to be that which comes as close as it can to the over-liberal canons of common sense without restoring the contradictions. But the more closely we approach this ideal point of liberality, the more risk we run of subtly reinstating a contradiction for posterity to discover.'[1] (An earlier version of *M.L.* was found to be inconsistent.)

To sum up this chapter, I have tried to explain in a manner adequate for our purposes in the critical considerations which are to follow, how the Leibniz-Frege-Russell programme of the logicist philosophy of mathematics has been carried out. It results in the actual construction of (interpreted) mathematical systems. Each of them consists on the one hand of postulates and rules of inference by which can be derived (i) all truth-functional tautologies, (ii) the unobjectionable theorems of the theory of classes or sets and (iii) the theory of quantification; and, on the other hand, of postulates for the avoidance of inconsistency. We have been granting, for the sake of philosophical argument, the claim that the formalisms possess (or could, after purely technical modifications, be made to possess) the required deductive power and freedom from contradiction—this last even though expert mathematical opinion is still divided on the point.

Do these formalisms support the philosophical thesis to the effect that pure mathematics is part of logic? And does the logicist philosophy of mathematics give a satisfactory account of applied mathematics? To deal with these questions will be our next task.

[1] *M.L.*, p. 166.

III

MATHEMATICS AS LOGIC: CRITICISM

AMONG the problems to which any philosophy of mathematics must address itself are, as was pointed out at the outset of this essay, first that of the structure and function of pure mathematics, second that of the structure and function of applied mathematics and thirdly the problems centering round the concept of infinity. The logicist answer to the first of these questions can, in accordance with the preceding chapter, be illustrated by its account of the proposition '$1+1=2$', which is roughly this:

Following Frege and Russell the number 1 is defined as a property or, more usually, as a class, namely the class of all those classes each of which contains a single element. More precisely, a class x contains only one element, in other words x is a member of the class of classes 1, $(x \in 1)$, if (i) there exists an entity, say u, such that $(u \in x)$ and if (ii) for any two entities v and w, if $(v \in x)$ and $(w \in x)$ then $v=w$. (Indeed, if any two entities which are elements of x are identical, they are one entity and, as such, in x.) The number '2' is analogously defined. It is explained that the phrase 'y is a member of the class of classes 2', $(y \in 2)$, holds if (i) there exists an entity, say u_1, such that $(u_1 \in y)$ and another entity u_2, such that $(u_2 \in y)$ and if (ii) for any entity, say v, if $(v \in y)$ then $(v=u_1) \lor (v=u_2)$.

In terms of these definitions of 1 and of 2, we can now express '$1+1=2$' firstly in terms of the logic of truth-functions; secondly in terms of the logic of quantification, from which we need the notion of the universal quantifier; and, of course, thirdly in terms of the logic of classes, from which we need the notions of the sum-class $\alpha \cup \beta$ (which is the class of elements which are members of either α or β) and of the product-class $\alpha \cap \beta$ (which, we remember, is the class of all elements common to α and β). If, in particular, $(\alpha \cap \beta) = \wedge$, *i.e.* if the product-class is the null-class—then α and β have no members in common.

To simplify our semi-formal definition we assume that x and y, which occur in it, are not empty and have no common member. We define: '$1 + 1 = 2$' by '$(x)(y)(((x \in 1) \& (y \in 1)) \equiv ((x \cup y) \in 2))$'. In words, on the assumption that x and y are non-empty classes with no common member, for any classes x and y—if x is an element of 1 and y is an element of 1—then, and only then, is their logical sum an element of 2. We might say that this definition has, among other things, reduced the addition of numbers to the class-theoretical operation of forming the sum-class of two classes.

If we can accept the logicist account of pure mathematics as illustrated by its analysis of '$1 + 1 = 2$', then the logicist account of applied mathematics as illustrated by its analysis of '1 apple and 1 apple make two apples' presents no further difficulty. We are then dealing, simply, with two statements of logic. If a and b are two classes of apples (not empty and with no element in common) then the above formula becomes for them $((a \in 1) \& (b \in 1)) \equiv ((a \cup b) \in 2)$. In other words '$1 + 1 = 2$' is a statement of logic about classes of classes in general, while '1 apple and 1 apple make 2 apples' is a statement of logic about classes of classes in particular—not an empirical statement about a world in which there happen to be physical apples with certain characteristics. Indeed what is logically true of classes of classes in general is logically true of classes of classes of apples, pears, numbers, etc.

Logicism knows no separate problem of pure and applied geometrical propositions and of their mutual relations. It arithmetizes, to use Weierstrass' and Felix Klein's expression, the whole of geometry after the fashion of the Cartesian analytic geometry; and thus incorporates geometry into the logicist system. It is clear that the validity of this approach depends entirely on whether we can accept the logicist account of pure and applied arithmetic.

As regards the third of the problems which any philosophy of mathematics has to face, the concept of mathematical infinity, the logicist account of the sequence of natural numbers involves the assumption of actual infinities. But although logicism, following Cantor, employs this notion most liberally by developing a mathematics of infinities of various sizes and various internal structures, its mathematical theory is not backed by any philosophical theory or analysis.

These preliminary remarks suggest the following order as appropriate to an attempt at assessing the logicist philosophy of mathematics. I propose first of all to argue that although logicism claims to reduce mathematics to logic, it does not at all clearly demarcate the field of logic. Next I shall try to show that the logicist account of pure and applied mathematics does not do justice to the fact that whereas the propositions of pure mathematics, *e.g.* that '$1+1=2$', are *a priori* or non-empirical, those of applied mathematics, *e.g.* '1 apple and 1 apple make two apples', are *a posteriori* or empirical. In brief, I shall argue that the fundamental difference between non-empirical and empirical concepts and propositions is ignored. Next I shall consider the logicist use of the notion of actual infinity, which is involved in the mathematical concept of natural number but not in the empirical. I shall try to show here that this use raises questions which logicism does not in any way answer. Lastly I shall give some attention to the logicists' account of pure and applied geometry, both for its own sake, and because it underlines and drives further home the objections laid against their analysis of pure and applied arithmetic.

1. *The logicist account of logic*

The logic, to which logicism holds pure mathematics to be reducible, presupposes the fundamental dichotomy of all knowledge into empirical and non-empirical or, as it has become customary to express it since the time of Kant, into *a posteriori* and *a priori*. This dichotomy is accepted by philosophers belonging to an old and broad stream of tradition including Plato, Aristotle, Leibniz, Hume, Kant, Frege and Russell. It is rejected by Hegel, by modern absolute idealists such as Bradley and Bosanquet, and by pragmatists of various persuasions.

The dichotomy is explained in different ways, but all similar in intention to the one which follows here, which is sufficient for our present purpose. We shall here assume that one understands what is meant by a statement's describing a possible perception or sense-experience; and by one statement's logically implying or entailing another. We can then say (in almost Kantian fashion) that a statement is *a posteriori* if, and only if, it (i) describes a possible sense-experience or (ii) if it is internally consistent and entails a statement describing a possible sense-experience. Thus 'the paper on which this book is printed is white' is *a posteriori* because of its describing a sense-perception. And 'all books are printed on white paper'—whether true or false—is an *a posteriori* statement since it entails, *e.g.*, that the paper on which this book is printed is white.

A proposition which is not *a posteriori* is *a priori*. Examples of *a priori* propositions are: $p \vee \sim p$, or any other truth-functional tautology; $1+1=2$, or any other proposition of pure mathematics; and perhaps 'man has an immortal soul' and other such assertions of theology. (The question whether such theological propositions are 'meaningless' need not occupy us since our having to dispense with them could, at most, only deprive us of a few convenient examples.) To the distinction between *a posteriori* and *a priori* propositions there corresponds a similar distinction between *a posteriori* and *a priori* concepts. A concept is *a posteriori* if in applying it to an object one is stating an *a posteriori* proposition.

Philosophers who dichotomise all knowledge in the manner just indicated into *a priori* and *a posteriori*, all seem—with the possible exception of Mill—to regard the propositions both of logic and of pure mathematics as *a priori*. What is controversial among them is, whether, within the class of *a priori* propositions, a further distinction is to be made between those of logic and those of pure mathematics. The fields of pure mathematics and of rational theology share the characteristic of being *a priori*; yet the two can be clearly distinguished and neither of them is reducible to the other. It is similarly possible that pure mathematics shares its *a priori* character with logic, and that they yet are irreducible the one to the other.

We have seen in chapter I that Leibniz's logicism was based on a clear but too narrow conception of logical propositions or, as he called them, truths of reason; and on a clear conception of formal demonstration or proof. The latter conception has been further clarified and perfected by his successors. The conception of logical propositions has, on the contrary, become more and more blurred—and through this the whole thesis that pure mathematics is deducible from logic has suffered from an ineluctable obscurity.

In order to see this, assume that there is a characteristic of propositions, say L, which some possess and others lack, and which is possessed also by all such propositions as are deducible from premisses possessing it. A characteristic which fulfils this requirement (of being 'hereditary') would be, for example, truth; one which would not fulfil it, might be, for example, triviality. We need not assume that every proposition is clearly characterized by possessing or lacking L. There may be border-line cases.

Now, in its original form logicism assumed that there is a hereditary characteristic L of logical propositions, namely their logical character, and it set out to show that (i) certain propositions, say $l_1, l_2, \ldots, l_n$, do among others obviously possess L; and (ii) from these all the propositions of pure mathematics, among them, say, $m_1, m_2, \ldots, m_n$, can be formally deduced—in a sense which we leave undiscussed and uncriticized for the moment. The propositions of pure mathematics, therefore, also possess L. Two distinct claims, it will be observed, are here being made. The mathematical claim of logicism is to have deduced the propositions *m* from the propositions *l*. The philosophical claim is to have exhibited clearly that the propositions *l*, and therefore the propositions *m*, have the general characteristic *L*. The vindication of the philosophical claim presupposes that the mathematical claim can be vindicated. But the mathematical claim, of the deducibility of the propositions *m* from the propositions $l_1, \ldots, l_n$, can be vindicated without exhibiting the $l_1, \ldots, l_n$ to possess a common general characteristic.

If we look at Quine's logicist system, for example, we 'find no assertion, much less any argument, to the effect that the premisses of the logico-mathematical system possess a *general characteristic L* which, as the result of deductions and definitions, though not *prima facie*, is seen to be possessed by the propositions of pure arithmetic. The premisses are merely enumerated. They are members of a list, and not obvious possessors of a general characteristic *L*. Quine's system attempts and, as we may assume, fulfils a mathematical task; but it does not at all support the logicist thesis that pure mathematics is reducible to logic, since it does not pretend to have explained the notion of a proposition of logic. It has been argued that certain other logicist systems are preferable to Quine's, but in none of them is the list of postulates supplemented by a general characterization of them as logical. The lack of this characterization is recognized by most contemporary logicists and those near to them.[1]

But the obscurity of the logicist account of logic is due only partly to the fact that it cannot show the premisses of any satisfactory logistic system to possess a general characteristic *I*. This incapacity might after all be explained, or explained away, by saying that *L* is like yellowness (or, according to some, like moral goodness) in being an unanalysable characteristic; and that its possession by the axioms of a given logicist system is, fortunately, apprehended immediately. This indeed seems to have been Frege's position before the discovery of the class-theoretical antinomies. Since then, every logicist system has had to include at least one postulate whose acceptance has to be justified on purely pragmatic grounds. Russell would not claim for the vicious-circle principle, and its supplementary assumptions, that it had the immediately-obvious and intuitively-undeniable character of a principle of logic; and neither would Quine claim this for his more elegant version. The logicist account of logic is philosophically inadequate beyond its mere obscurity.

If we assume that it is incapable of substantial improvement;

[1] See, *e.g.*, Carnap, *Introduction to Semantics*, Harvard, 1946, chapter C.

the following alternatives suggest themselves among others. First, logic and mathematics may not be one *a priori* science but two separate *a priori* sciences. It is possible, in other words, to characterize a large class of *a priori* propositions, including those of the traditional logic and many propositions of *Principia Mathematica*, by a general hereditary characteristic *L*, and to characterize a large class of *a priori* propositions, including those of pure arithmetic and many other propositions of pure mathematics, by a general hereditary characteristic *M*. But no subset of the possessors of *L* contains the premisses from which all the propositions of pure mathematics follow. This is in fact the view which logicism principally set out to refute. A variant of it was held by Kant and it still shows its influence in the mathematical philosophies of both formalism and intuitionism.

Secondly, it might be held that the impossibility of finding a general characteristic *L* obviously possessed by the axioms of a logicist system and either obviously or demonstrably possessed by its theorems, shows that logic and pure mathematics are connected even more intimately than this. On this view logic and pure mathematics would be so much one science that even to make a *prima facie* distinction between them, as do Frege and Russell, would be impossible. To speak of a reduction of mathematics to logic would then be just as pointless as to speak of a reduction of logic to logic or of mathematics to mathematics. If this view were correct it should be possible to find a general characteristic, say *A*, obviously possessed by the axioms of a logico-mathematical system and either obviously or demonstrably by its theorems. The search, however, for such a general characteristic (of *analyticity*, as it has been called, in defiance of older terminologies) as would cover both logic and mathematics has, so far, been unsuccessful—which is perhaps the less surprising, in view of the pragmatic principles included among the postulates of logicist systems, especially principles which are almost indistinguishable from empirical hypotheses about the universe.

A third conceivable view would assert the impossibility, not only of finding a characteristic *L*, but of finding a general characteristic *A* which distinguished the propositions of logic and mathematics on the one hand from empirical propositions on the other. According to this view the unity of logic and mathematics would be based on the impossibility of any sharp distinction even between *a priori* and *a posteriori* propositions. It is strange that this, the very antithesis of logicism, should be held by, of all philosophers, Quine, whose main aim as a logician has been to perfect the system of *Principia Mathematica*. According to this point of view the programme of deducing pure mathematics from logic is replaced by that of showing how very many different propositions can be deduced from very few. The alleged logical difference between empirical propositions and the non-empirical propositions of logic and mathematics is regarded as merely a pragmatic difference, a difference in the degree of tenacity with which various thinkers hold on to various propositions—the propositions of logic and mathematics being those which are least easily, the empirical those which are most easily, dropped. The original logicism of Frege and Russell becomes a thorough-going pragmatic logicism. In this compound name 'logicism' expresses no more than a pious historical memory.

I shall later argue that mathematical propositions and theories are exact in a sense in which empirical propositions and theories are not; and that mathematical theories are existential in a sense in which—in many senses of 'logic'—logic is not. That is to say that I shall, on the whole, be arguing in favour of the view that mathematics and logic are two separate *a priori* sciences.

2. *The logicist conflation of empirical and non-empirical concepts*

The Frege-Russell definition of natural numbers, and of the concept of a natural number, is rightly regarded as having been one of the most impressive features of logicism. There is, indeed, a difference of opinion, as we have seen, between those who accept Frege's account of numbers as independent entities, and those who follow Russell in regarding the words for number-concepts as incomplete symbols—symbols only contextually defined. The main point, however, of the Russell-Frege account is not thereby affected. The point was to have asserted the definability of the notion (its definability in purely logical terms) and to have offered a definition; whether on realist or nominalist metaphysical principles did not matter.

The logicist analysis has been attacked on various grounds. It has been objected, *e.g.*, that the analysis is circular. Since having a certain number, as a property of a class, is defined in terms of the notion of similarity between classes, the question arises as to how we establish the similarity. Apparently, in some cases at least, we must count; that is to say, apply the concept of number. Frege foresaw this objection, and urged that his definition of the number of a class in terms of the similarity of classes is neither more nor less circular than the usual definition of the direction of a straight line in terms of the parallelism of straight lines. Frege and Russell do, however, make implausible assumptions. They are not, indeed, compelled to hold that the similarity or lack of it between two classes, *i.e.* the presence or absence of a one-one correspondence between their members, can be established in every case. But they do assume it to be true of any two classes; either that they are or else that they are not similar, even if there is no possible way of finding out. The nature of this assumption is, at best, obscure and in need of justification.[1]

Another, and possibly an even more widely urged, objection to the Russellian definition of number is that a purely logical concept cannot be defined by reference to a non-logical hypothesis. Russell, as we have seen, does lay himself open to this objection. He not only *defines* every natural number n as having a unique successor $n+1$, but has to *assume as a non-logical hypothesis* the axiom of infinity, the axiom which 'assures us (whether truly or falsely) that there are classes having n members and thus enables us to assert that n is not equal to $n+1$'. Without this axiom, he continues, 'we should be left with the possibility that n and $n+1$ might both be the null-class'.[2] This form of objection to Russell's definition of number—that it violates the logicist programme—is justified as far as it goes. The programme was to reduce mathematics to logic and not to logic *plus* non-logical hypotheses. But this objection does not go far enough.

Let us consider a concept 'n is a Natural Number' which is so defined that it does *not* entail 'n has a unique immediate successor'. In other words, we admit the possibility, envisaged by Russell, that the number-sequence comes to an end. Moreover, if there is a last Natural Number we assume it to be so great that no one—whether scientist or grocer—needs to be perturbed about it. The concept of a Natural Number is certainly applicable to groups of perceptual objects. The statement, for example, that the group of apples on this table has the Natural Number 2 is an application of the concept 'Natural Number'; and the truth of the statement is independent of whether the Natural Numbers form an endless sequence or not.

Let us consider next the concept 'n is a natural number' which is so defined that it *does* entail 'n has a unique immediate successor', and, therefore, 'n has infinitely many successors'. *This* concept may not be applicable to groups of perceptual objects. Indeed the truth, for example, of the statement that the group of apples on this table has the natural number 2 is dependent on the natural numbers forming an endless sequence of which 2 is by definition a member. If they did

---

[1] See also Waismann, *Einführung in das mathematische Denken*, Vienna, 1947, pp. 76 ff.; English translation, New York, 1951.
[2] *Introduction to Mathematical Philosophy*, p. 132.

not 'happen to form' such a sequence, the concept 'natural number' would be empirically empty. The concepts 'Natural Number' and 'natural number' thus differ not only in logical content, *i.e.* in their logical relations to at least one other concept, namely that of having a unique immediate successor; but possibly also in their range or extension.

Moreover, the hypothesis of the infinite sequence of natural numbers by which the concept 'natural number' is defined and provided with its infinite range admits of no empirical falsification or confirmation. It leaves room for further 'hypotheses' of a similar kind, one of which 'assures us' that the class of natural numbers is completely given, another that in addition the class of its subclasses is also given. But there are also hypotheses which assure us to the contrary. This freedom of defining mutually inconsistent concepts *and of providing them by the definition* with different ranges shows that none of these concepts is empirical. The Natural Numbers on the other hand are empirical concepts, characteristics of perceptual patterns, such as groups of strokes or of temporally separated experiences. They and their relations to each other are found, not postulated.

Again, the Natural Numbers 1, 2, etc. are inexact in the sense that they admit of border-line cases, *i.e.* patterns to which they can with equal correctness be assigned or refused. They share this inexactness with other empirical concepts. The natural numbers 1, 2, etc. on the other hand are exact.

In applying pure mathematics, we 'interpret' not only pure number-concepts in terms of Natural Numbers, but also pure mathematical relations and operations (such as addition) in terms of empirical relations and operations. The difference between physical or empirical concepts, and the corresponding mathematical ones of different logical content and range of reference, is generally acknowledged by applied mathematicians, in particular by those who are looking for new mathematical models of experience. For example, the following are some introductory remarks prefacing an attempt to mathematicize certain parts of economics in a new way.[1] 'In all these cases where such a "natural" operation is given a name which is reminiscent of a mathematical operation—like the instances of "addition" above— one must carefully avoid misunderstandings. This nomenclature is not intended as a claim that the two operations with the same name are identical—this is manifestly not the case; it only expresses the opinion that they possess similar traits, and the hope that some corres-

[1] v. Neumann and Morgenstern, *Theory of Games and Economic Behaviour*, 2nd edition, Princeton, 1947, p. 21.

spondence between them will ultimately be established.' Similar remark on the correspondence between mathematical and empirical concepts and relations, with similar warnings, are found in a standard treatise on a (fairly) new mathematicizing of statistics in terms of measure-theory.[1] That the application of mathematics to experience pre-supposes a correspondence between empirical concepts and those 'idealizations' of them which are mathematical concepts—between, e.g., displacements or velocities on the one hand and vectors on the other—is almost a commonplace. My contention is that this is so even in the case of 'Natural Number' and 'natural number'.

But the reasons for separating 'Natural Number' from various concepts of 'natural number', and for separating other empirical concepts from 'corresponding' mathematical concepts, have still not been completely expressed. In comparing concepts in respect of their logical content, we have tacitly allowed two assumptions: first that it is always clear whether or not one concept stands in a certain logical relation to another; and secondly that the logical relations which can hold between mathematical concepts are not essentially different from those that can hold between empirical. Both these assumptions are mistaken.

As regards the first, it would be generally agreed that the logical relations which connect mathematical concepts, especially in forma-lized systems, are much more precisely defined than the logical relations between empirical concepts. A consequence of this is that the question whether or not two mathematical concepts stand in a certain logical relation admits of decision in cases where the question regarding the corresponding empirical concepts does not. The making precise of the intuitive notion of entailment and of other logical relations can be, and has been, achieved in a number of different ways. The logical network between the mathematical concepts is dependent on the logical system—in particular the logical formalism—in which it happens to be embedded. Empirical concepts are not so anchored in any similar system.

As regards the second assumption, I shall argue later that the logical relations in which it is possible for empirical concepts to stand differ in a fundamental way from those that can hold between mathematical concepts. This difference will be shown to be connected with the inexactness of the former and the exactness of the latter.

My aim in this section has been to show that the logicist account of applied mathematics implies an illegitimate conflation of mathe-matical number-concepts and of corresponding empirical ones.

[1] H. Cramér, *Mathematical Methods of Statistics*, Princeton, 1946, pp. 145 ff.

Ignoring the difference between the corresponding concepts, logicism cannot, and does not, say anything about the nature of this correspondence. This is a task which, eventually, will have to be taken up. (See chapter VIII.)

3. *The logicist theory of mathematical infinity*

It has been clear ever since Greek times that if one allows oneself to think in terms of actual infinities, the totality of dimensionless points which lie on or constitute a line-segment, and the totality of dimensionless moments which lie in or constitute a stretch of time, are *in some sense* greater than the totality of all positive integers or the totality of all fractions. The attempt to understand continuous spatial configurations, and continuous temporal changes, in terms of numerical relations—the attempt to arithmeticize geometry and chronometry—seems indeed to require us to compare infinite classes, with respect to their numerical size and ordinal structure. It has been conjectured that the Greeks' rejection of actual infinities, as expressed in particular by Aristotle, prevented them from unifying arithmetic and geometry after the fashion of Descartes, Leibniz and their successors. This unification has naturally and almost unavoidably led to a mathematics which distinguishes between the sizes of various actual infinities, and between their structures, and which calculates with infinite cardinal and ordinal numbers.

The historical importance of the 'naive' transfinite mathematics, created by Cantor and incorporated almost wholly into *Principia Mathematica*, can hardly be overestimated; for without it the less naive theories would have had little to analyse, to be critical of, to reconstruct. In what follows I shall be sketching briefly some of the central notions of this transfinite arithmetic, their importance for the understanding of continuous shapes and processes, and the features of them which seem to call for a reconstruction.[1]

Let us call a class x a proper subclass of y if every member of x is a member of y while not every member of y is a member of x. It is obviously impossible in the case of a finite class, *e.g.* {1, 2, 3}, to establish a one-one correspondence between it and any proper subclass of it, *e.g.* {1, 2}. At least one member of the class would always remain unmatched. This is not so in the case of infinite classes. Here a one-one correspondence between the class and a proper subclass of it can be established. For example, the infinite class of all the natural num-

---

[1] In addition to the books by Fraenkel, mentioned above, the reader will find an excellent and not too technical introduction in E. V. Huntington's *The Continuum*, 2nd edition, Harvard University Press, 1917, Dover Publications, 1955.

bers has the class of all those that are even, as one of its proper subclasses. Here a one-one correspondence between class and proper subclass can be established by adhering to the following rule: (a) Put the natural numbers into their order of magnitude, 1, 2, 3, . . . and put the even numbers into *their* natural order 2, 4, 6, . . . and (b) match the first number of the first sequence with the first number of the second, the second number of the first with the second of the second, and so on. Every member of the first will then have one and only one partner in the second, and no member of either will be unpartnered.

In terms of the notions of 'proper subclass' and 'similarity' the distinction between finite and infinite classes can be sharply defined. An infinite class is such a class as can be put into one-one correspondence with one of its proper subclasses. A class which is not infinite is finite. Clearly the Frege-Russell definition of number, or more precisely of cardinal number, covers also transfinite cardinals. Thus the transfinite cardinal number a is defined as the class of all classes similar to the class {1, 2, 3, . . .}—the class of all natural numbers. A class with cardinal number a is also called 'denumerable'. It can easily be shown (*vide* the relevant text-books) that the class of all rational numbers and the wider class of all (complex) algebraic numbers (those numbers which are roots of polynomial equations with integers as coefficients) are denumerable. It is similarly easy to show that the classes of all rational numbers, and of all algebraic numbers, which lie between any two of them are denumerable.

The number-concept in terms of which modern mathematical analysis, in particular the differential and integral calculus, has been developed is the concept of a real number. It is in connection with this concept that the idea of the actual infinite has become problematic, not only to philosophers of mathematics but even to pure mathematicians themselves. The class of all real numbers greater than 0 and, say, equal to or smaller than 1 is non-denumerable, *i.e.* is not similar to any class of cardinal number a. The proof of this has been given by Cantor and is very roughly this: every real number in the interval can be represented by a decimal fraction of form $0 \cdot a_1 a_2 a_3 \ldots$ which does not terminate (rational numbers will be periodic in this representation).[1] Assume now that, if possible, all these decimal fractions are written out as a sequence, *i.e.* in one-one correspondence to the sequence 1, 2, 3, ... Now replace the first number of the first decimal fraction, the second number of the second, the third number of the third, etc. by different numbers, *e.g.* by stipulating that each of these

---

[1] A brief account of the classical concept of 'real number' is given in Appendix A.

numbers be replaced by 1 if it is not itself 1, otherwise by 2. The number so created is obviously a decimal fraction but one which does *not* occur in the sequence of fractions: for it differs from the first decimal fraction in the first place, from the second decimal in the second place and so on. There is thus no one-one correspondence between the class of all real numbers and the class of all integers. It can be shown that the classes of real numbers in any interval are similar. All these similar classes and others similar to them have the same cardinal number c, which is the cardinal number of the continuum.

$\mathfrak{a}$ and c are thus two different transfinite numbers and $\mathfrak{a}$ is smaller than c in the precise sense that while $\mathfrak{a}$ can be put in one-one correspondence with a proper subclass of c, it cannot be brought in one-one correspondence with c itself. Are there any cardinal numbers greater than c? According to Cantor and Russell there are—even more than enough of them. The following consideration will give some idea of Cantor's argument. Consider the class {1, 2, 3} and form the class of all its subclasses including the null-class and the class itself. The new class will then be {∧, {1}, {2}, {3}, {1, 2}, {1, 3}, {2, 3}, {1, 2, 3}}. The original class has 3 members, the class of its subclasses 2^3. Cantor argues that given any finite or infinite class with cardinal number x there exists the class of all its subclasses with cardinal number 2^x so that whereas any class of cardinal number 2^x has a subclass which is similar to any class of cardinal number x the converse does not hold. For every x there exists thus the greater 2^x and there is no greatest transfinite cardinal number.

We have seen that transfinite cardinal numbers are in some cases equal to one another; and that among cardinal numbers one may be in a precise sense greater than another. If a and b are *finite* numbers, the following three relations may hold between them: $a = b$, $a > b$, $a < b$. If a and b are *transfinite*, it is *prima facie* not inconceivable that they are not comparable with each other. In order to establish the same kind of comparability between transfinite as holds between finite cardinal numbers, set-theorists had to assume that every class can be put into a certain standard order—even if no effective method is known for achieving this. The assumption is that every class can be well ordered, *i.e.* put in an order which fulfils the following conditions.[1] (i) There exists a relation R such that (a) if x and y are distinct elements of the class, then either $(x \text{ R } y)$ or $(y \text{ R } x)$; (b) if $(x \text{ R } y)$ then x and y are distinct; (c) if $(x \text{ R } y)$ and $(y \text{ R } z)$ then $(x \text{ R } z)$. (ii) Any subclass of the series has a first member. (For this is by no means necessary. The

---

[1] See Huntington, *op. cit.*

series, *e.g.* of real numbers between 0 and 1 excluding 0 and arranged in order of magnitude, has no first member.) The postulate that every class can be well ordered is relevant not only to logic and to the arithmetic of transfinite numbers, but also to 'ordinary' mathematics such as the theory of the Lebesgue integral.

The postulate that every class can be well-ordered links the transfinite arithmetic of cardinal numbers with the transfinite arithmetic of ordinal numbers, which also form an unlimited hierarchy and are defined in terms of the one-one correspondence between classes as ordered by various relations. Some of the notions defined in this theory are of great importance in topology and other branches of pure mathematics. There is not much point in adding to this very sketchy outline of transfinite cardinal arithmetic an outline of ordinal arithmetic which would have to be similarly sketchy. What has to be said about the former applies on the whole also to the latter.

The transfinite mathematics, of the nature and grandeur of whose content the preceding remarks may have conveyed some slight hint, was quickly discovered to lead to contradictions. As we have seen, the theory permits the making of statements about *all* members of finite and infinite classes of any cardinal number, *e.g.*, about the class of *all* natural numbers, the greater class of all subclasses of that class, the still greater class of all subclasses of the class just mentioned, etc. But if we assume the existence of the class of *all* cardinal numbers then this assumption, which is not forbidden by Cantor's theory, is incompatible with its theorem that there is no greatest transfinite cardinal number. The class of all cardinal numbers cannot be conceived as completely given.

The importance of this antinomy, both for Cantor's theory and for the logicist version of it, is well described by the author of a standard work on Cantor's theory.[1] 'What is disquieting about this antinomy is', he says, 'not that a contradiction arises, but that one was not prepared for a contradiction: the class of all cardinal numbers seems *a priori* just as unsuspicious as the class of all natural numbers. Hence originates the insecurity, as to whether perhaps other infinite classes, possibly all of them, are not such pseudo-classes affected with contradictions ... and then the task to eliminate this insecurity. ...'

The principles on which, in logicist formalisms, in particular in *Principia Mathematica*, the antinomy of the greatest cardinal, together with the antinomy of the class of all those classes which do not

---

[1] F. Hausdorff, *Mengenlehre*, 3rd edition, p. 34; available also in Dover Publications.

contain themselves as members, and other antinomies are avoided are, unfortunately, principles which are neither obviously nor demonstrably logical in any accepted sense of the term. They have, and are generally agreed to have, the character of *ad hoc* remedies. Those who propose them do not claim to have diagnosed the source of the malady for which they are prescribing but merely express the hope that contradictions will thus be avoided.

Now if a concept, such as the concept of actually given infinite totalities of different cardinal number, can be rendered harmless only by *ad hoc* remedies and only provisionally, one may adopt towards such a concept any of several philosophical attitudes. One may try first of all to replace the defective concept by a sound one which serves the same purpose. This is what Hilbert and his school have attempted. These mathematical philosophers, as we shall see in greater detail at a later stage, require the statements of a mathematical theory to be clearly linked to (though not necessarily descriptive of) perceptible or constructible objects and perceptible operations upon these objects. The reason lies in the thesis that statements describing actual or possible perceptions can never be mutually inconsistent. The task for these philosophers and mathematicians is to replace the 'non-constructive' concepts of naive and logicist theories by 'constructive' ones. This task is specially important for the mathematics of real numbers which in classical mathematics are defined non-constructively, in terms of actually infinite classes (*e.g.* as infinite decimal fractions regarded as somehow completely 'written down' or otherwise spread out).

Another possible attitude is to jettison either all actual infinities or all non-denumerable ones, and to pay the price, by not only accepting, in some parts of mathematics, in particular analysis, greater complexity and prolixity, but also by sacrificing other parts of the subject. This is the attitude taken up by Brouwer and others who wholly or partly follow him in his effort to eliminate actual infinite totalities from mathematics.

On the whole Frege and Russell, in their analysing and, if we may use the expression, logicizing of arithmetic, have uncritically employed the Cantorian actual infinities. Just as those who uncritically use the concept of a physical object are 'naive' rather than philosophical realists, so the logicists who use the concept of actual infinities uncritically cannot claim to have a philosophy of infinity. That there should be this gap in it, is a serious count against their philosophy of mathematics.

4. *The logicist account of geometry*

Any known geometrical discipline can be developed in two fundamentally different ways. According to the one the geometric entities—points, lines, planes, etc.—are put into one-one correspondence (or are identified) with numbers or sets of numbers and the geometric relations are similarly put into similar correspondence with relations between numbers. This type of analytic or arithmeticized geometry presupposes a highly developed number-concept, in particular the notion of real numbers. If these are conceived after the fashion of Cantor and logicism, they presuppose in turn the notion of denumerable *and* non-denumerable actual infinities. Any doubts concerning actual infinities thus affect the legitimacy of absorbing geometry into arithmetic and mathematical analysis.

The other way of developing a geometric discipline is to consider the geometric entities, whether real or fictitious, and the relations between them, independently of all numerical representation. The geometric entities are now defined only partially, by stating their relations to other geometric entities of the same or different kind, and not also by such characteristics as would enable a person to perceive, construct or imagine them. When it is stated for example that through any *point* which does not lie on a given *straight line* only one *line* can be drawn which is *parallel* to the given *line*, the geometric system which contains this statement as a postulate or theorem does not contain as part of itself any statement which will help us to identify *points* or *lines* (whether *parallel* or not), either exactly or approximately, with marks on a blackboard or with any other such physical objects.

A logicist finds no fault, of course, with non-denumerable actual infinities, or with the concept of real number involving them, or with the arithmetization of all known geometry by means of this concept. He may, therefore, quite properly assert that the system of *Principia Mathematica* or any similar system has also 'provided for' geometry if, as Quine says, 'we think of geometrical notions as identified with algebraic ones through the correlations of analytic geometry'.[1]

To the logicist's account of geometry no objections are likely to arise, with which we have not already dealt, in considering his account of pure and applied arithmetic and his uncritical acceptance of actual infinities. Essentially new objections could only be based on some supposed impossibility of arithmetizing geometry at all, whether by the method of logicism or any other. So long as we are not prepared

[1] *Op. cit.*, p. 81.

to argue this, our arguments can only concern the means of arithmetization, *i.e.*, in the present case, the non-geometrical key concepts of logicism. I propose, nevertheless, to consider the logicist account of geometry in some detail. It will help to emphasize and reinforce some of the general arguments already considered, in particular those to the effect that logicism conflates empirical statements and concepts with non-empirical ones.

Whereas the distinction between natural numbers and Natural Numbers may appear to some people merely shocking or perverse, the parallel distinction between, say, Euclidean triangles and physical triangles is almost generally agreed. Hardly anybody would identify a Euclidean triangle—which we consider now apart from any numerical representation—with a drawn triangle, or regard any drawn triangle as an instance of the concept 'Euclidean triangle'. The distinction was very clearly made by Plato according to whom the participation of physical triangles in the Form of the mathematical triangle is quite different from instantiation. It is expressed again and again by philosophers and mathematicians, among them the great systematic geometer, Felix Klein. Klein says, for example: 'it is true in general that *fundamental concepts and axioms* (of geometry) *are not immediately facts of perception, but are appropriately selected idealizations of these facts*' (Klein's italics).[1]

As an example of a geometrical statement of special interest, let us take the familiar postulate of parallels, that for any straight line and any point not on it *there exists* one, and only one, parallel line through the point. It must be understood as one of a number of propositions—about geometrical points, lines, etc.—which together permit us to deduce the whole of Euclidean geometry, apart from any numerical representation. The following features of the geometrical proposition must be emphasized:

It is, first of all, an *a priori* proposition in the sense explained above (p. 54). There are no logical relations of deducibility or incompatibility between it on the one hand and perceptual statements on the other. Our proposition is logically disconnected from perceptual statements or, briefly, disconnected from perception. Indeed if perceptual statements describe and relate physical points and physical straight lines all of which have length, breadth and height, our statement is about objects which, whatever other characteristics they may have, are *not* three-dimensional, but are in the case of geometrical points dimensionless, in the case of lines one-dimensional.

---

[1] *Elementary Mathematics from an Advanced Standpoint*, Geometry, English translation, Dover Publications, p. 186.

It is important to note in this connection that the geometrical straight line is of infinite extension. As distinguished from perceptual line-segments a geometrical line-segment is embedded in an infinite line. Here as in the case of Natural Numbers, which are not embedded in an infinite sequence, and of natural numbers which are, the extrapolation of perception to infinity is a feature which distinguishes the geometrical from the corresponding empirical concept.

Secondly, the statement obviously admits of incompatible alternatives or, briefly, it is not unique. A proposition p—such as our postulate—is not unique if from its incompatibility with some other proposition, say q, it does *not* follow that one of the two incompatible propositions is false. Instances of a type of proposition which is not thus unique are rules. ('To smoke immediately after breakfast' is incompatible with 'Not to smoke before lunch', yet neither of these is false. Rules are neither true nor false.) On the other hand *a posteriori* propositions and logically necessary propositions are unique. (See chapter VIII below.)

That geometrical statements are not unique has been demonstrated by the construction of self-consistent non-Euclidean geometries. Neither the postulate of parallels nor its negation are confirmed or falsified by perceptual statements, in particular about perceptual space. What can be confirmed or falsified by perceptions—experiments and observations—is not a geometry or any set of *a priori* statements but a physical theory using the geometry. What was falsified by the Michelson-Morley experiment was not Euclidean geometry, but a physical theory using it. What is confirmed by experiment is not a particular non-Euclidean geometry but again a physical theory using it. Kant's thesis that Euclidean geometry is *the* geometry of perceptual space is just as mistaken as the thesis that the geometry of perceptual space is not Euclidean.

Thirdly, the postulate of parallels has a characteristic which it does not share with all other geometrical statements. It is existential in the sense that it goes beyond stating what a concept entails, by asserting it (or demanding or assuming it) not to be empty. It not only determines the concepts of a point or a line, leaving the question about their range to be decided by independent inquiry; it determines their range directly. The two questions whether, say, 'being a man' entails 'being mortal' and whether there exists a man are quite separate and different. In defining the term 'being a man', to take it as entailing 'there exists at least one man' would be rejected as doing more than determining its logical content, meaning, etc. This is just what the postulate of parallels does. It determines the range of 'parallel to a line' not merely indirectly by stating logical relations between different concepts, but determines it directly.

The nature of 'existential' statements in mathematics will occupy us later (chapter VIII). Here it is sufficient to point out that if a concept is 'defined' as applying to objects, which are not given in perception, it might be argued that such objects must be either found elsewhere or otherwise provided. This view is strongly supported by Hilbert and Bernays[1] who state quite clearly that in any axiomatic theory—and at present we are concerned with such a theory apart from its possible numerical representation and incorporation into a logicist formalism —'we are dealing with a fixed system of things (or with a number of such systems) which constitutes *a range of subjects, demarcated* from the outset, for all predicates from which the statements of the theory are formed'. The axioms, unlike ordinary definitions, provide the predicates whose logical content they determine, with particulars (subjects). (It is of some interest to note that the existential character of, *e.g.*, the postulate of parallels distinguishes it from rules, with which it shares the characteristics of being logically disconnected from perception and of not being unique.)

Fourthly the postulate of parallels is an idealization—it idealizes perceptual judgements. The notion of idealization stands in need of more explanation than is usually given to it. It does in particular require a characterization of what is idealized, what idealizes and the relation between them. For the present it will be sufficient to say that the postulate and other geometrical statements are *a priori* propositions which—although logically disconnected from perception—can yet for particular *purposes* be used interchangeably with empirical propositions. It is worth emphasizing that our fourth characteristic is not purely logical. It refers to a possible purpose which the proposition characterized may serve. This is as it should be. Whether a physicist, for example, uses the Euclidean idealization of perception or a non-Euclidean one depends precisely on the purpose in hand.

The logicist account of geometry consists, as we have seen, in the arithmetization of geometry—geometrical concepts being represented by ordered classes of numbers, their instances by the elements of these classes, and their relations by numerical ones. It is clear that the arithmetization of geometry and its consequent incorporation into logicist formalisms does in no way affect the difference between geometrical propositions and those empirical propositions of which the former are idealizations. Logicism as a philosophy of mathematics does

---

[1] *Op. cit.*, vol. 1, p. 2.

not account for the differences and relations between them. Moreover, even if we were to agree that philosophy should not consider pure geometry apart from its arithmetization, all the objections raised against the logicist conflation of natural numbers and Natural Numbers can be raised against the logicist account of geometry.

IV

MATHEMATICS AS THE SCIENCE OF
FORMAL SYSTEMS: EXPOSITION

WE turn now to another line of thought with another historical root. As Leibniz sought the source of the self-evidence and the content of mathematics in logical relations between propositions and concepts, so Kant sought it in perception. And, just as Leibniz conceived the guiding principles of logicism, so Kant was led to anticipate the guiding principles of two modern movements in the philosophy of mathematics: formalism and intuitionism.

For Kant the role of logic in mathematics is precisely the role it has in any other field of knowledge. He holds that in mathematics, although the theorems follow from the axioms *according to* principles of logic, the axioms and theorems are not *themselves* principles of logic, or any application of such principles. He regards them, on the contrary, as descriptive, namely as describing the structure of two perceptual data, space and time. Their structure manifests itself as something which we find in perception, when we abstract its varying empirical content. Thus in perceiving two apples, the iteration which is perceived is a feature of the space and time in which the apples are located. The same structure manifests itself further in our deliberate geometrical constructions, both in making such constructions possible and in confining them within limits—permitting the construction, for example, of three-dimensional objects but not of four-dimensional.

Hilbert, who in his practical programme adapted Kant's guiding idea, expresses Kant's 'fundamental philosophical position', and his own, in the following words: '... something which is presupposed in the making of logical inferences and in the carrying out of logical operations, is already given in representation (*Vorstellung*): *i.e.* certain extra-logical concrete objects, which are intuitively present as immediate experience, and underlie all thought. If logical thinking is to be secure, these objects must be capable of being exhaustively surveyed, in their parts; and the exhibition, the distinction, the succession of their parts, and their arrangement beside each other, must be given, with the objects themselves, as something that cannot be reduced to anything else or indeed be in any need of such reduction.'[1]

Hilbert shares this fundamental position with Brouwer and his school as well as with Kant. If mathematics is to be restricted—entirely and without qualification—to the description of concrete objects of a certain kind, and logical relations between such descriptions, then no inconsistencies can arise within it: precise descriptions of concrete objects are always mutually compatible. In particular, in this kind of mathematics, there are no antinomies to trouble us, generated by the notion of actual infinity; and for the simplest of reasons, namely that the concept of actual infinity does not describe any concrete object.

Yet—and here is the root of the disagreement between formalists such as Hilbert and intuitionists like Brouwer—Hilbert does not think his position requires him to abandon Cantor's transfinite mathematics. The task he sets himself is the accommodating of transfinite mathematics within a mathematics conceived, in Kantian fashion, as concerned with concrete objects. 'No one will ever be able to expel us', he says, 'from the paradise which Cantor has created for us.'

His way of reconciling concrete, finite mathematics with the abstract and transfinite theory of Cantor is something Hilbert again owes—at least fundamentally—to Kant.[2] It was not, indeed, in the philosophy of mathematics that Kant employed the principle on which Hilbert's reconciliation proceeds. Kant employed it in a part of philosophy which for him was much more important—the reconciliation of moral freedom and religious faith with natural necessity. Arguing in this context, Kant first pointed out that the notion of moral freedom (and some other notions, including that of actual infinity) were Ideas of Reason which were unrelated to perception, in the sense of being neither abstracted from it nor applicable to it. He then argued that any system containing notions applicable primarily to concrete objects (such as the mathematics and physics of his day) could indeed be amplified by Ideas, but only provided the amplified system could be shown to be consistent. Proving consistency, within a system embracing both the findings of theoretical science on the one hand and, on the other, the Ideas of morals and faith, was Kant's way as he himself put it 'of making room for faith'.

In quite similar fashion Hilbert distinguishes between the concrete

---

[1] Hilbert, *Die Grundlagen der Mathematik*, Sem. der Hamburger Universität, vol. 6, p. 65. Also Becker, p. 371.

[2] See, *e.g.*, *op. cit.*, p. 71.

or real notions of finite mathematics and the ideal notions (Ideas) of transfinite mathematics. In order to justify the adjunction of ideal notions to the real, he too requires a proof that the system is consistent. Hilbert's task is thus to prove the consistency of a system comprising finite and transfinite mathematics. He adopts the Kantian theses (i) that mathematics includes descriptions of concrete objects and constructions and (ii) that the adjunction of ideal elements to a theory requires a proof of the consistency of the system thus amplified. In his hands these have been transformed into what is claimed to be a practical programme for founding mathematics upon what is perceived or perceivable. We have now to examine this.

1. *The programme*

To show that a system of propositions—*e.g.* the theorems of a mathematical theory—is internally consistent is to show that it does not contain two propositions one of which is the negation of the other or a proposition from which any other proposition would follow. (The second formulation also holds for systems in which negation is not available.) Only in the case of very simple systems is it possible to compile a list of all their propositions and to check the list for inconsistency. In general, a more complex investigation into the structure of the system as a whole will be necessary.

Such an investigation presupposes that the system is clearly demarcated and capable of being surveyed. The demarcation, as Frege saw, is secured to some extent by axiomatization: *i.e.* by listing the undefined concepts in the system, the presupposed assumptions in it, and lastly, the inference-rules (the rules for deducing theorems —from the assumptions and already deduced theorems). We have mentioned (in chapter II above) various axiomatizations of the logic of propositions, of classes, and of quantification. Similar axiomatizations have often been given for other systems, such, *e.g.*, as (unarithmetized) geometry and parts of theoretical physics. Axiomatization may be more or less strict, depending on the extent to which the rules of sentence-formation and of inferential procedure are more or less explicitly and precisely formulated.

For proving the consistency of a system two methods are available: the direct and the indirect. In some cases it can be shown by combinatorial means that inconsistent statements are not deducible in a given theory. In other cases the direct method proceeds by exhibiting a perceptual model of the theory. More precisely it consists (i) in identifying the objects of the theory with concrete objects, (ii) in identifying the postulates with exact descriptions of these objects and their mutual relations, and (iii) in showing that an inference within the system will not lead to any other than exact descriptions. Since mathematics abounds in concepts of actual infinities which cannot be identified with perceptual objects, the use of the direct method is restricted to certain small parts of mathematics.[1]

A theory involving actual infinities can—at least *prima facie*—be tested for consistency only by the indirect method. One proceeds in this by establishing a one-one correspondence between (a) the postulates and theorems of the original theory and (b) all or some of the postulates and theorems of a second theory, which is assumed to be consistent. The consistency of this theory can in some cases be reduced to a third one. But none of these theories can have a concrete model.

Amongst indirect proofs of the consistency of any geometrical or physical theory the most common are based on arithmetization, *i.e.* on representing the objects of these theories by real numbers or systems of such. This is by no means surprising. For on the one hand the creative work of mathematicians, at least since Descartes, has been characterized by the demand that all mathematics should be capable of being embedded in arithmetic; and, on the other hand, the creative work of physicists, at least since Galileo, has been characterized by the demand that all physics should be mathematized. These are philosophical demands and convictions and they have led to extensions of mathematics so as to make it capable of accommodating all physical formalisms; and they have led to such extensions of arithmetic as to make it capable—by the use of one-one correspondences—of accommodating all mathematics, in particular all geometry and abstract algebra. It cannot indeed be said *a priori* that this arithmetization of science has no limits. But the reducibility to arithmetic of physical and mathematical theories which contain ideal notions, and which cannot be proved consistent by the direct method, raises the question of the consistency of arithmetic itself. Before Hilbert, no practical programme for proving the consistency of arithmetic had been suggested. (If mathematics should be found reducible to an obviously consistent logic, this problem would not, of course, arise.)

And Hilbert's basic idea, here, is as ingenious as it is simple. The mathematician deals with concrete objects or systems of such. He can therefore rely on 'finite methods'; in other words he can rest content with the employment of concepts which can be instantiated in perception, with statements in which these concepts are correctly applied, and with inferences from statements of this type to other such statements.

[1] See, *e.g.*, Hilbert-Bernays, *op. cit.*, p. 12.

76 THE PHILOSOPHY OF MATHEMATICS

Finite methods do not lead to inconsistencies, especially in mathematics where the concrete objects can be effectively demarcated.

Classical arithmetic does, of course, deal with such abstract and ideal objects as actual infinities. But even when on this account non-finite methods have to be used *within* arithmetic it may nevertheless be possible to regard or reconstruct arithmetic *itself* as a concrete object which can be dealt with by finite methods. It would be natural to expect this concrete object to possess properties capable of throwing light on classical arithmetic as usually conceived. It may in particular be expected to have a property the possession of which would guarantee the consistency of the classical arithmetic.

Before attempting a more detailed exposition of these points one can hardly do better than formulate the programme for proving the consistency of the classical arithmetic in Hilbert's own words: 'Consider the essence and method of the ordinary finite theory of numbers: This can certainly be developed through number-construction by means of concrete, intuitive (*inhaltlicher, anschaulicher*) considerations. But the science of mathematics is in no way exhausted by number-equations and is not entirely reducible to such. Yet one can assert that it is an apparatus which in its application to whole numbers must always yield correct numerical equations. But then there arises the demand to inquire into the structure of the apparatus to an extent sufficient for the truth of the assertion to be recognized. And here we have at our disposal, as an aid, that same concrete (*konkret inhaltliche*) manner of contemplation, and finite attitude of thinking, which had been applied in the development of the theory of numbers itself for the derivation of numerical equations. This scientific demand can indeed be fulfilled, *i.e.* it is possible to achieve in a purely intuitive and finite manner—just as is the case with the truths of the theory of numbers—those insights which guarantee the reliability of the mathematical apparatus.'[1]

The consistency of the classical arithmetic—including, we may say, the main parts of Cantor's theory—is to be proved and the programme would appear to be (i) to define with all possible clarity what is meant in mathematics by finite methods as opposed to non-finite, (ii) to reconstruct as much as possible of classical arithmetic as a precisely demarcated concrete object which is given to, or realizable in, perception and (iii) to show that this object has a property which clearly guarantees the consistency of classical arithmetic.

The formalist not only needs the assurance that his formalism formalizes a consistent theory, but also that it completely formalizes

[1] *Op. cit.*, p. 71; Becker, p. 372.

what it is meant to formalize. A formalism is complete, if every formula which—in accordance with its intended interpretation—is provable within the formalism, embodies a true proposition, and if, conversely, every true proposition is embodied in a provable formula. (This is the original meaning of the term 'completeness' which has also other, though related, meanings in the literature some of which have no reference to an original, non-formalized, theory.) For some such formalisms there are available mechanical methods—decision procedures—by which one can decide for any formula whether it is provable or not and whether consequently the embodied proposition is true or false. The ideal would be a consistent, complete and mechanically decidable formalism for all mathematics.

2. *Finite methods and infinite totalities*

Incompatibility is a relationship between propositions or concepts. Perceivable objects and processes cannot be incompatible with each other. Again, propositions cannot be incompatible with each other if they *precisely* describe such objects and processes; for a description implying incompatibility between entities that cannot be incompatible could not be precise. Yet the trouble is that there is no general test for deciding whether a description is or is not precise. Attempts such as Russell's sense-data theory to mark out in general objects which can be precisely described—or such attempts as are made by theories like Neurath's theory of 'protocol sentences' to mark out propositions which are precisely descriptive—are by no means universally accepted as successful. In mathematics it seems to be otherwise. Here it seems comparatively easy to demarcate a narrow field of perceptual objects and processes which will be capable of precise description, or at least of a description free from contradictions. In the elementary theory of numbers we deal with such objects and processes. The methods of dealing with them, the so-called finite (or 'finitary') methods, are explained in the above mentioned papers by Hilbert and in the classic *Die Grundlagen der Mathematik* by Hilbert and Bernays.[1] Consistently with these texts the point of view might be put as follows.

The subject matter of the elementary theory of numbers consists of the signs '1', '11', '111', etc., *plus* the process of producing these signs by starting with '1' and putting always another stroke beyond the last stroke of the previous sign. The initial figure '1' and the production-rule together provide the objects of the theory; these objects can be abbreviated by use of the ordinary notation, the numeral '111', *e.g.*, being written as '3'. The small letters a, b, c, etc.

[1] See also Kleene's *Introduction to Metamathematics*, Amsterdam, 1952.

are employed to designate unspecified figures. For operations performed on the figures one uses further signs: brackets, the sign ' ≡ ' (to indicate that two figures have the same structure) and the sign '<' (to indicate that one figure is in an obvious and perceivable way contained in another). Thus $11 < 111$, *i.e.* if beginning with '1' we build up '11' and '111' by parallel steps the former will be finished before the latter.

Within this elementary theory of numbers, one can perform and describe concrete addition, subtraction, multiplication and division. The associative, commutative and distributive laws, and the principle of induction are nothing else than obvious features of these operations. Thus '$11 + 111 = 111 + 11$' is an instance of '$a + b = b + a$', an equation which asserts in a general way that the production of figures by iterating the stroke does not depend on order.

Again the principle of induction, the most characteristic of all the principles of arithmetic, is, in the words of Hilbert and Bernays [1] not an 'independent principle' but 'a consequence which we take from the concrete construction (*Aufbau*) of the figures'. Indeed if (a) '1' has a certain property and (b) if, provided the property is possessed by any stroke-expression, it is also possessed by the succeeding stroke-expression (the expression formed by putting a further '1' after the original) then this property will be seen to be possessed by any stroke-expression that can be produced. Having defined the concrete fundamental operations by means of the concrete principle of induction, one can define the notion of prime numbers, and construct for any given prime number a bigger prime number. The process of recursive definition can also be defined and performed concretely. For example the factorial function $\rho (n) = 1.2.3 \ldots n$ is recursively defined by (a) $\rho(1) = 1$ and (b) $\rho(n + 1) = \rho(n).(n + 1)$. This definition prescribes in an obvious way how, beginning with $\rho(1)$, and using nothing but concrete addition and multiplication, we can build up $\rho(n)$ for any perceptually given figure n.

Elementary arithmetic is the paradigm of mathematical theory. It is an apparatus which produces formulae, and which can be entirely developed by finite methods. This statement, however, the meaning of which has just been illustrated from the development of elementary arithmetic, is still needlessly imprecise, and requires an actual and explicit characterization of what is to be meant by 'finite methods'.

First, every mathematical concept or characteristic must be such that its possession or non-possession by any object can be decided by inspection of either the actually constructed object or the constructive

[1] *Op. cit.*, p. 23.

process which would produce the object. The second of these alternatives introduces a certain latitude in determining finite characteristics and the finite methods consisting in their employment. Thus one is reasonably content with a process of construction which is 'in principle' performable. Indeed it is at this point, namely when the choice arises between making the formalist programme less strict or sacrificing it, that some relaxation of the finite point of view may be expected.

Secondly, a truly universal proposition—a proposition about all stroke-expressions for example—is not finite: no totality of an unlimited number of objects can be made available for inspection, either in fact or 'in principle'. It is, however, permissible to interpret any such statement as being about each constructed object. Thus, that all numbers divisible by four are divisible by two means that if one constructs an object divisible by four, this object will have the property of being divisible by two. Clearly this assertion does not imply that the class of all numbers divisible by four is actually and completely available.

Thirdly, a truly existential proposition—to the effect, *e.g.*, that there exists a stroke-expression with a certain property—is equally not finite: we cannot go through *all* stroke-expression (of a certain kind) to find one which has the property in question. But we may regard an existential proposition as an incomplete statement to be supplemented by an indication either of a concrete object which possesses the property or of the constructive process yielding such an object. In the words of Hermann Weyl,[1] an existential proposition is 'merely a document indicating the presence of a treasure without disclosing its location'. Propositions which involve both universal and existential assertions—*e.g.* to the effect that *there exists* an object which stands to *every* object in a certain relation—can again only be suffered as *façons de parler* promising the exhibition of perceivable or constructible relationships.

Fourthly, the law of excluded middle is not universally valid. In finitist mathematics one permits neither the statement that all stroke-expressions possess a property *P* nor the statement that there *exists* a stroke-expression which does not possess *P*—unless these statements are backed by an actual construction. One consequently cannot admit as universally valid the unqualified disjunction of these two statements, that is to say the law of excluded middle.

Even in elementary arithmetic there is occasion for using, in a restricted way, transfinite methods, in particular the principle of

[1] *Philosophy of Mathematics and Natural Science*, Princeton, 1949, p. 51.

excluded middle. But whereas transfinite methods here are easily replaceable by finite ones quite sufficient for their perceivable or constructible subject matter, the situation is different, as we have seen already at various stages of the argument, in analysis. This fundamental difference between elementary arithmetic and analysis in its classical form is due—as has frequently been pointed out—to the fact that the central notion of analysis, that of a real number, is defined in terms of actual infinite totalities. (See Appendix A.)

We have seen that every real number between 0 and 1 (we can disregard the real numbers outside this interval without loss of generality) can be represented by a decimal fraction of the form $0 \cdot a_1 a_2 a_3 \ldots$ where the dots indicate that the number of decimal places is a, *i.e.* denumerably infinite. If the numbers to the right of the decimal point do not terminate, *i.e.* if they are not from a certain place onwards all zeros, and if their sequence shows no periodicity, then the infinite decimal fraction represents an irrational number. Every place of the decimal fraction can be occupied by one of the numbers 0 to 9. The totality of these possibilities, which represents the totality of all real numbers in any interval is, we have seen, greater than the totality of all integers and greater than the totality of all rational numbers. Its cardinal number c is greater than a, the cardinal number of any denumerable set.

In order to appreciate the nature of this statement about real numbers it will be well to consider the representation of real numbers by binary fractions of the form $0 \cdot b_1 b_2 b_3 \ldots$. Here, just as the first place to the right of the decimal point indicates tenths, the second hundredths, the third thousandths and so on, so the first place to the right of the binary point indicates halves, the second quarters, the third eighths, etc. Again, just as every place of a decimal fraction can be occupied by any number from 0 to 9 inclusive, so every place of a binary fraction—every b—is occupied by either 0 or 1. Moreover just as all real numbers can be represented by all decimal fractions, so all real numbers can be represented by all binary fractions—the choice of the decimal, the binary or any other system being a purely external matter.

Assume now that all natural numbers are given in their natural order and in their totality thus: 1, 2, 3, 4, 5, 6, ... Now form a finite or infinite subclass from the totality, indicating *the choice* of a number for the subclass by writing 1 in its place, and indicating the *rejection* of a number by writing in its place 0. If we choose 2, 4, 5, . . . and reject 1, 3, 6, we shall thus write 010110 . . . . It is clear that every infinite sequence of zeros and ones determines one and only one subclass of the class of natural numbers in their natural order. But we have just seen that every infinite sequence of zeros and ones determines one and only one real number between 0 and 1 (in the binary representation). There is thus a one-one correspondence between the class of all subclasses of natural numbers and the class of all real numbers between 0 and 1 and, as can be easily shown, the class of all real numbers in any interval. In speaking of a real number the classical analyst is committed to the assumption that it is 'possible' to pick out a subclass from the *actual* totality of all natural numbers. In speaking of all real numbers he is not only committed to assuming the actual totality of all natural numbers but also the *greater actual* infinite totality of all subclasses of this class (see p. 63). The assumption of such totalities implied in speaking of a real number, or even of all real numbers, transcends the finite point of view and the employment of finite methods.

Classical analysis transcends the finite point of view not only by assuming actual infinite totalities, but by using the law of excluded middle without qualification. If not all members of a class have a certain property P then at least one member has the property *not-P* and *vice-versa*—indifferently whether the class in question be finite, denumerably infinite or greater than these. Another non-constructive principle of classical analysis and the theory of sets was made explicit by Zermelo. This is the so-called principle or axiom of choice (*Auswahlprinzip*). Hilbert and Bernays formulate it as follows:[1] 'If to every object x of a genus $\mathfrak{G}_1$ there exists at least one object y of genus $\mathfrak{G}_2$, which stands to x in the relation $B(x, y)$, then there exists a function ϕ, which correlates with every object x of genus $\mathfrak{G}_1$, a unique object $\phi(x)$ of genus $\mathfrak{G}_2$ such that this object stands in the relation $B(x, \phi(x))$ to x.'

Another way of expressing the axiom of choice is to say that given a class of classes, each of which has at least one member, there always exists a selector-function which selects one member from each of these classes. (One might 'picture' the selector-function as a man with as many hands as there are non-empty classes—picking out one element from each of them.) It is obviously possible to exhibit a selector-function for a class consisting of a finite number of finite classes. When it comes, however, to picking out one member from each of an infinite number of finite classes, still more from an infinite number of infinite classes, the exhibition of the selector-function, as a feature of perceivable or constructible objects or processes, is clearly out of the question. That the axiom of choice is implicitly assumed in a great deal of analysis and set-theory only became clear to mathematicians

[1] *Op. cit.*, p. 41.

after Zermelo discovered it to have been a tacit assumption in the proof that every class can be well-ordered, and that in consequence the cardinal numbers of any two (finite or infinite) classes are comparable (see p. 64).[1]

Thus, on Hilbert's showing, classical mathematics has as its hard core a perceivable, or at least in principle perceptually constructible, subject-matter, to which fictitious, imperceivable and perceptually non-constructible objects, in particular various infinite totalities, are adjoined. To this adjunction of 'fictitious' subject-matter there correspond (i) ideal concepts which are characteristic of it—*e.g.* Cantor's actual infinities, and transfinite cardinal and ordinal numbers—(ii) ideal statements describing either it or operations upon it—*e.g.* the unqualified law of excluded middle, or the axiom of choice—(iii) ideal inferences leading either from statements of finite mathematics to ideal statements or from ideal statements to other ideal statements.

This adjunction of ideal concepts, statements and inferences to a theory is, of course, not at all new in mathematics. Thus in projective geometry it has proved of great use to introduce an ideal point at infinity on every straight line and to define it as the point at which all lines parallel to the given line intersect; and to introduce, in every plane, an ideal line containing all the points at infinity of all the lines in the plane. There can, of course, be no question of 'the ideal point common to two parallel lines' denoting any perceptually-given or constructible entity; the reasons for demanding points of intersection of parallel lines require any set of parallel lines to have *one* point of intersection, *not two* points of intersection, one, as it were, at each end of the parallel lines.[2] By adjoining ideal points, lines and planes to the 'real' ones, one creates concepts which, although logically related to the concepts to which they have been adjoined, are even less characteristic of perception than the former. Even if 'real point' and 'real line' can *cum grano salis* be said to describe perceptual objects, no amount of salt will make it plausible to say that 'ideal point' and 'ideal line' are perceptual characteristics.

The introduction of ideal elements into projective geometry, into the algebraic theory of numbers and mathematical theories in general, has, according to Hilbert, been one of the glories of creative mathe-

---

[1] As to the use of the axiom in topology, in the theory of Lebesgue measure, etc., see J. B. Rosser, *Logic for Mathematicians*, New York, 1953, pp. 510 ff.

[2] For an explanation of the reasons for the introduction of ideal points, lines and planes and for further details see, *e.g.*, Courant and Robbins, *What is Mathematics?*, Oxford, 1941, and later editions, especially chapter IV.

matical thinking. The emergence of antinomies as a result of this adjoining of infinite totalities to elementary arithmetic requires according to him not their abandonment but some proof that an extended arithmetic—the combination into one system of finite and transfinite objects and methods—is free from contradiction. How this is to be achieved is, he argues, suggested by considering elementary arithmetic.

His crucial point here is that elementary arithmetic can be conceived of in two different ways; on the one hand, quite naturally, as being a *theory about* the regulated activity of constructing stroke-expressions, and, on the other hand, somewhat artificially, as being a *formalism, i.e.* as itself a regulated activity of constructing perceptual objects—this time, of course, not stroke-expressions but formulae. The arithmetical theory consists of statements, the arithmetical formalism of symbol-manipulations and their results. The formalism can, just like the regulated activity of constructing stroke-expressions, become the subject-matter of another theory, usually called a 'meta-theory'. We are thus led to distinguish between two kinds of constructing activities—stroke-construction and formula-construction; and between two kinds of theory—the original theory about stroke-construction and the new 'metatheory' about formula-construction.

The connection between arithmetical theory, arithmetical formalism and metatheory about the arithmetical formalism is obviously quite intimate. In its broad outlines it is founded on the fact that the *same* physical objects, *e.g.* $\langle 1+1=2 \rangle$ or $\langle 1+1=3 \rangle$ (the objects between the French quotes) function in distinct though corresponding ways, in the arithmetical theory and in the arithmetical formalism. The formalism may be built up in such a manner that it becomes possible to distinguish among its rules two kinds in particular: (a) rules for the production of such formulae as correspond (like our two examples) to statements of the theory and which we shall call statement-formulae; (b) rules for the production of such as (like the first example, but unlike the second) correspond to true statements or theorems of the theory and which we shall call theorem-formulae.

In asserting that a certain physical object is, in the context of the formalism, a statement-formula or a theorem-formula, we are speaking *about* formula-construction and are making a statement of meta-theory. This statement is *finite*, in that it asserts of a perceptual object, or of the process by which it is produced, a purely perceptual or (literally!) formal characteristic. The *formal* characteristic of a statement-formula's being a theorem-formula corresponds to the *logical* characteristic of a statement's being a theorem.

To this correspondence between the formal characteristics of the formalism and the logical characteristics of the theory, others can be added. Perhaps the most important of these is the correspondence between the formal consistency of the formalism and the logical consistency of the theory. To assert that the *theory* is logically consistent is to assert that not every statement of the theory is also a theorem of the theory. (This definition, as has been indicated before, has the advantage of avoiding the use of the notion of negation.) To assert that the *formalism* is formally consistent is to assert that not every statement-formula of the formalism is also a theorem-formula. In view of the correspondence (mediated by their embodiment in the same physical objects) between statement-formulae and theorem-formulae on the one hand, and statements and theorems on the other, we are entitled to say that to demonstrate formal consistency is at the same time to demonstrate logical consistency.

We now turn to non-elementary arithmetic. The subject-matter of this arithmetical *theory* is, of course, no longer finite. But it may be possible to construct an arithmetical *formalism*—with statement-formulae and theorem-formulae corresponding as before to statements and theorems of the theory; and this formalism could then be the subject-matter of a metatheory. Since the subject-matter, namely formula-construction, would be finite, the metatheory would be just as finite as elementary arithmetic, from which it would differ only by being about a different kind of perceptual construction. If a formalism corresponding, in the required manner, to the theory of non-elementary arithmetic can be constructed, then we can again, by demonstrating *formal* consistency of the formalism, *eo ipso* establish *logical* consistency of the theory. Indeed we can do this by strictly *finite* methods, since our subject matter—the regulated activity of formula-construction—is perceptual, or at least in principle perceptually constructible. Our next task, therefore, must be to consider the formula-constructing activities, or formalisms—both formalisms considered by themselves and formalisms which are at the same time formalizations of theories.

3. *Formal systems and formalizations*

Once a formal system has been constructed a new 'entity' has been brought into the world—a system of rules for the production of formulae. These formulae are perceptual objects which can be distinguished and classified by means of perceptual characteristics which are possessed either by the formulae themselves or by the process of their production, in particular by the sequence of formulae which successively lead from an initial formula to the formula under consideration. In a formal argument we must ignore any correspondence between the formal properties of the formal system and the logical properties of any pre-existing theory, even though to establish such a correspondence was the guiding motive in constructing the formal system.

According to Hilbert the content of mathematics is still propositions; in the case of elementary arithmetic they are propositions about stroke-expressions and their production, in the case of the amplified (classical) arithmetic they include in addition propositions 'about' ideal objects. The formal systems which he constructs are merely means by which, in virtue of the correspondence between formal and logical properties, he studies the pre-existing mathematical theories. His formalisms are formalizations.

Yet since no insight derived from the pre-existing theory is permitted to enter the arguments concerning the formal system; since, that is to say, from the point of view of these arguments, no theory needs to exist of which the formal theory is a formalization, the possibility is opened to us of regarding the formal theory not merely as an instrument for investigating a pre-existing system of propositions, but as the subject-matter of mathematics itself. There are good grounds for this. On the one hand, there is no reason why the subject-matter of metamathematics should not be extended to any kind of formal manipulation of marks. On the other hand a phenomenalist philosopher, or one of a similar philosophical persuasion, might well—for philosophical reasons of a general kind—deny the existence of ideal propositions and thus declare, *e.g.*, the amplified arithmetic with its ideal objects and propositions to be meaningless or simply false. If so, he would, with H. B. Curry[1] propose to define mathematics as 'the science of formal systems'. In other words, whereas to Hilbert a formal system is the Leibnizian 'thread of Ariadne' leading him through the labyrinth of mathematical propositions and theories, the *strict formalist* regards mathematics as having this thread—and nothing more—for its subject-matter.

The change from Hilbert's formalist point of view to the strict formalism of Curry leaves the former's mathematical results untouched. It represents, however, a transition to a different philosophical point of view. Mathematics has now no truck *with anything but* formal systems, in particular not with ideal, non-perceptual entities. Hilbert's position is analogous to that of a moderate phenomenalist who would admit physical-object concepts as auxiliary

[1] *Outlines of a Formalist Philosophy of Mathematics*, Amsterdam, 1951.

—if fictitious—concepts, in terms of which sense-data would be ordered or purely phenomenalistic statements made—even if physical-object concepts could not be 'reduced' to sense-data, or to purely phenomenalist concepts. Strict formalism on the other hand is analogous to a phenomenalism which would admit only sense-data and purely phenomenalist statements.

Strict formalism as a philosophy of mathematics is nearer than Hilbert's view to Kant's doctrine in the *Transendental Aesthetic*. According to Kant a statement in pure mathematics has constructions for its subject-matter—constructions in space and time, which by the very nature of these intuitions are restricted. According to strict formalism the subject-matter of mathematics is constructions, the possibility of which is restricted by the limits under which perception is possible; and our statements about these constructions are *demonstrationes ad oculos*, read off, as it were, from perception. They are true synthetic statements. However, their self-evidence is neither that of logical tautologies, nor, as Kant held, that attaching to supposedly *a priori* particulars. It is the self-evidence of very simple phenomenalist or sense-data statements. Statements about mathematical constructions are in other words empirical statements involving the least possible risk of error. This is the reason why in discussing the process of proof—one of the principal subjects of the science of formalisms—Curry says, very naturally, that it is 'difficult to imagine a process more clear cut and objective'.

For Hilbert the *raison d'être* of formal systems is to save and safeguard the pre-existing—albeit somewhat modified—classical theories, in particular Cantor's theory of sets. For Curry formal systems are the substitutes of classical mathematics. From these fundamental differences between moderate and strict formalism others follow. For Hilbert, who intends to establish the (logical) consistency of theories *via* the (formal) consistency of formal systems, a (formally) inconsistent formal system is useless. Not so for Curry. He maintains that for the acceptability or usefulness of a formal system 'a proof of consistency is neither necessary nor sufficient'.[1] Indeed inconsistent formal systems, he argues, have in the past proved of the greatest importance, *e.g.* to physics.

Both Hilbert and Curry deny the possibility of deducing mathematics from logic. Yet whereas Hilbert regards principles of reasoning which are sufficient for elementary arithmetic as logical principles of a finite and, as it were, minimal logic, Curry separates logic and mathematics even more drastically. It all hinges, he says,[2] 'on

---

[1] *Op. cit.*, p. 61.   [2] *Op. cit.*, p. 65.

what one means by "logic"—"mathematics" we have already defined. ... On the one hand logic is that branch of philosophy in which we discuss the nature and criteria of reasoning; in this sense let us call it logic (1). On the other hand in the study of logic (1) we may construct formal systems having an application therein; such systems and some others we often call "logics". We thus have two-valued, three-valued, modal, Brouwerian, etc. "logics", some of which are connected with logic (1) only indirectly. The study of these systems I shall call logic (2). The first point regarding the connection of mathematics and logic is that mathematics is independent of logic (1). ... Whether or not there are *a priori* principles of reasoning in logic (1), we at least do not need them for mathematics.'

Hilbert has never explicitly and at any length dealt with the philosophical problem of applied mathematics. He seems to favour the view that there is a partial isomorphism between pure mathematics and the realm of experience to which it is applied. Elementary arithmetic, that is to say, either is itself the empirical subject-matter of our study—a 'physics' of stroke-symbols and stroke-operations—or else can be brought into one-one correspondence with some other empirical subject-matter; for example, to take a trivial case, apples and apple-operations. The non-elementary parts of the amplified arithmetic, on the other hand, have no empirical correlates. Their purpose is to complete, systematize and safeguard the elementary core which alone either is empirical or has empirical correlates.

According to Curry, who is quite explicit on this question, we must distinguish between the *truth* of a formula within a formal system—*i.e.* the statement that it is derivable within the system—and the *acceptability* of the system as a whole. The former is 'an objective matter about which we can all agree; while the latter may involve extraneous considerations'.[1] Thus he holds that 'the acceptability of classical analysis for the purposes of application in physics is ... established on pragmatic grounds and neither the question of intuitive evidence nor that of a consistency proof has any bearing on this matter. The primary criterion of acceptability is empirical; and the most important considerations are adequacy and simplicity.'[2] When it comes to the application of mathematics Curry is a pragmatist. He does not go so far as the pragmatic logicist whose view of pure mathematics is also pragmatist and who denies that logical, mathematical and empirical propositions can be distinguished by any sharp criteria. (See p. 57.) The domain of formal theories and the propositions about their formal properties are, Curry holds, clearly demarcated.

[1] *Op. cit.*, p. 60.   [2] *Op. cit.*, p. 62.

Before describing some formal systems in outline, we may perhaps be allowed an imprecise, metaphorical characterization of the basic ideas of formalism. According to most philosophers, from Plato to Frege, the truths of mathematics exist (or 'subsist') independently of their being known and independently of their embodiments in sentences or formulae, even if these are needed for the truths to be grasped. It was Hilbert's ingenious programme—foreshadowed to some extent by Leibniz—so to embody the truths of classical mathematics that the perceptual features of the bodies or of the processes by which they are produced correspond to logical features of mathematical propositions. The theorem-formulae are, as it were, the bodies and the disembodied truths the souls—every soul having at least one body. This programme, as will be explained a little more precisely later, cannot be carried out. It has been demonstrated by Gödel that every embodiment of classical mathematics in a formalism must be incomplete; there are always mathematical truths which are not embodied in theorem-formulae.

In order to appreciate this result we must be a little more specific about the nature of formalisms. Hilbert remarks on a kind of pre-established harmony which favours the progress of mathematics and the natural sciences. Results which are achieved in the pursuit of quite diverse purposes often provide the much needed instrument for a new scientific aim. The logical apparatus of *Principia Mathematica*, which, on the basis of previous researches with still different aims, was devised for the purpose of reducing mathematics to logic, provided, in Hilbert's own particular case, the *almost* finished tool for executing his quite different programme. Where *Principia Mathematica* falls short is in its incomplete formalization. It is not wholly a system of rules for manipulating marks and formulae, in particular theorem-formulae in total independence of the fact that they can be interpreted as propositions of classical mathematics. But *Principia Mathematica* is an *almost* perfect foundation for the rigorous formalization of classical mathematics.

Indeed, of the formal systems, those outlined in discussing the logicist philosophy of mathematics are as good examples as any. This applies in particular to the propositional calculus and the formal system of Boolean class-logic. Here we shall do no more than describe the general nature of formal systems. They are machines for the production of physical objects of various kinds, machines whose properties have been made the subject of extensive and detailed inquiries by Hilbert, Bernays, Post, Carnap, Quine, Church, Turing, Kleene and many others. As the result of the work done by these authors the terms 'machine' and 'mechanical properties' have in logical contexts long ceased to be metaphorical. (Indeed, most important insights into the nature of formalisms, that is to say the most important theorems of metamathematics or, as it is also called, proof-theory, can most simply and clearly be formulated as statements to the effect that certain formula-producing machines can, and certain others cannot, be constructed.)

Strict formalism regards, as we have seen, all mathematics as the science of formal systems, whether they are formally consistent or not, and whether or not they are intended to be formalizations of pre-existing theories; and it has made the nature of formalisms *per se* easier to grasp. To do this has become necessary for any philosophy of mathematics. For there can be no doubt that whatever else mathematics may mean, either now or in the future, it must always include the science of formal systems.

A very clear characterization of formal systems in general is given by Curry.[1] Each is defined by a set of conventions, its so-called primitive frame. By indicating the primitive frame we are providing an engineer with all the data he needs (apart from his knowledge of engineering) for constructing the required formula-producing machine. Curry distinguishes the following features in any primitive frame:

(i) Terms

These are (a) *Tokens*, which are specified by giving a list of objects of different types, *e.g.* marks on paper, stones or other physical objects. (b) *Operations*, *i.e.* modes of combination for forming new terms. (c) *Rules of formation* specifying how new terms are to be constructed. For example, if marbles and boxes are among our terms and the enclosing of marbles in boxes among our operations, we might adopt the rule of formation permitting the enclosure of each marble in a box, and stipulate that the enclosed marbles belong to the same kind of term as the loose ones.

(ii) Elementary Propositions

These are specified by giving a list of 'predicates' with the number and kind of 'arguments' for each. For example, we may specify as predicates pieces of wood with n holes into which both enclosed and loose marbles can be fitted and then determine that our elementary propositions are all those pieces of wood the holes of which have been duly filled in by enclosed or loose marbles.

---

[1] *Op. cit.*, chapter IV.

(iii) Elementary Theorems

(a) *Axioms*, *i.e.* elementary 'propositions' which are stated to be 'true' unconditionally. (b) *Rules of Procedure* which are of the following form: 'If $P_1, P_2, \ldots, P_m$ are elementary theorems subject to such and such conditions, and if Q is an elementary proposition having such and such a relation to $P_1, P_2, \ldots, P_m$, then Q is true.' For example, if two pieces of wood with holes filled in by marbles are elementary theorems, then any piece of wood which is produced from the former by gluing them together is also 'true'.

In order to be able to speak of the primitive frame we must have names for the tokens, operations and predicates and also indications of the way in which the predicates are applied to terms. Specification of the features which constitute the primitive frame of a formal system must be effective or *definite* (a term used by Carnap). This means that it must be possible to determine after a finite number of steps whether an object has or has not this feature. Indeed if a formal system is to be capable of being treated by finite methods (à la Hilbert), if in other words what is to be proved about it can be proved by demonstrations *ad oculos*, then the properties of being a formal predicate, of being a formal axiom, of a formula's being formally derived from another in accordance with a rule of procedure, must all be definite.

The property of being a theorem-formula may be but it need not be definite; but the formal relation between a formula and the sequence of formulae constituting its proof must, of course, be definite. In most mathematical theories a formula does, so to speak, not bear on its forehead the mark of being a theorem, but the proof of it, once given, must be capable of being checked in a finite number of steps.

Many formal systems have been constructed by mathematicians in the present century. The motive of the activity has usually been the need so to embody propositions into formulae that the formal properties and relations *of the formulae* guarantee corresponding logical properties and relations of the propositions. Indeed, as we have seen, the ultimate purpose of Hilbert's programme, and what would be its consummation, is a proof of the logical consistency of the main body of classical mathematics reached *via* a proof of the formal consistency of a suitable formal system.

As has often happened before in other branches of mathematics, the study of formal systems led to unexpected results, to new problems, new techniques and to at least one new branch of pure mathematics, namely the theory of recursive functions. The importance of this theory is considered by the experts very great. Thus E. L. Post who has not only made important contributions to this subject, but who also has expressed its main ideas in a manner which makes them accessible to non-experts, expresses the view that the formulation of the notion of recursive functions 'may play a role in the history of combinatory mathematics second to that of the formulation of natural number'.[1]

The reader of a book on the philosophy of mathematics cannot expect that a full knowledge of these new ideas and techniques will be conveyed in it. Yet he will readily see that the question how far the correspondence between pre-existing theory and formal system can be established is of great philosophical relevance; and he will expect a report of results achieved by the mathematicians. *Prima facie* the complete embodiment of mathematical theories in formalisms may seem possible; and then it will at least be arguable that the pre-existing theories are merely 'intuitive' in the somewhat disparaging sense in which the term is used by mathematicians on the first few pages of their treatises before they get down to business, and that the said theories are merely heuristic preliminaries for the construction of formalisms and statements about them.

We must, therefore, attempt to give an account of some results in the science of formal systems, trusting the mathematicians—as we have always done so far—to have done their job efficiently.

4. *Some results of metamathematics*

Only a very brief and very rough outline of Gödel's main result and of some new developments connected with it can be given.[2] Suppression of 'technicalities' must here inevitably mean suppression of essential arguments and insights. To whet the appetite of the reader without crass misstatements is perhaps the best that can be done.

We assume with Hilbert that the method and results of elementary arithmetic (see p. 77) need no justification; and we consider a presumably consistent formal system F which is sufficiently expressive to permit the formalization of elementary arithmetic in it. This implies the

---

[1] *Bulletin of the American Mathematical Society*, 1944, vol. 50, no. 5.

[2] The fundamental paper is Gödel's ' Über formal unentscheidbare Sätze der Principia Mathematica und verwandter Systeme, I ' in *Monatshefte für Mathematik und Physik*, 1931, vol. 38. For 'an informal exposition of Gödel's theorem and Church's theorem' see J. B. Rosser's article of this title, *Journal of Symbolic Logic*, 1939, vol. IV, no. 2. An informal and formal account of Gödel's theory is found in *Sentences Undecidable in Formalized Arithmetic* by Mostowski, Amsterdam, 1952; also in Kleene, *op. cit.*, and Hilbert-Bernays, *op. cit.*, vol. 2. The theory of recursive functions is developed from first principles and without a specialized logical symbolism in R. Péter's *Rekursive Funktionen*, 2nd ed., Budapest, 1958. For an excellent general survey of the present state of the theory, see John Myhill, *Philosophy in Mid-Century*, Florence, 1958.

requirement that all arithmetical expressions correspond to formal expressions in such a fashion that no formal theorem of F corresponds to a false arithmetical proposition. If a formal statement, say f, is the formalization of an arithmetical proposition a, a is also said to be an (arithmetical) interpretation of, or the intuitive meaning of, f.

Let us say that F *completely* formalizes elementary arithmetic provided that in the case of every formal statement f which is the formalization of an arithmetical statement either f or $\sim f$ is a formal theorem of F; or briefly, provided that f is decidable. Hilbert aimed at the complete formalization of (substantially) the whole of classical mathematics. Gödel has shown that even a formal system which formalizes no more than elementary arithmetic does *not* formalize it completely.

The incompleteness of F is established by the actual construction of a formal statement f which formalizes an arithmetical proposition while yet neither f nor $\sim f$ is a formal theorem of F, *i.e.* while f is undecidable. The interpretation of f reminds one of the liar-paradox: 'The proposition which I am now asserting is false.' If the assertion of the proposition is correct then the proposition is false, from which it follows that the assertion is incorrect. The statement is 'about' itself. It states its own falsehood, and states no more. It is this kind of self-reference which Gödel's formal proposition possesses. But whereas in the liar-paradox the relation between linguistic expression and its meaning is far from clear, Gödel's formal proposition is as clear as F and arithmetic.

We now turn to the construction of the undecidable f (following Mostowski's exposition). Since F formalizes elementary arithmetic, the integers and properties of integers must have formal counterparts in F. The formal integers or numerals will be printed in bold-faced type so that, *e.g.*, **1** corresponds to 1. The formal properties of integers will be expressed by $W(.)$, different formal properties being distinguished by different subscripts. If $W_0(.)$ is the formal counterpart of 'x is a prime number', then $W_0(\mathbf{5})$ is the formal counterpart of the arithmetical proposition that 5 is a prime number. The set of all formal properties of integers can be ordered in many ways into a sequence and we consider one of these sequences, say,

$$(1) \quad W_1(.), \ W_2(.), \ W_3(.), \ldots$$

In order now to construct the self-referring formal proposition let us formulate first any formal proposition arrived at by 'saturating' some formal property with the numeral corresponding to its subscript. Such formal propositions are $W_1(\mathbf{1})$, $W_2(\mathbf{2})$, $W_3(\mathbf{3})$, ... We next pick out, say, $W_5(5)$. This formal proposition may or may not be a formal theorem of F. Let us assume that it is not, *i.e.* that

$$W_5(5) \text{ is not a formal theorem of } F.$$

This proposition is on the face of it not a formal proposition of F, but is a real proposition about a formal proposition, namely about the formal proposition $W_5(5)$. It is in Hilbert's sense a metastatement, belonging to the metalanguage in which we talk about F. Similarly the property:

(2) $W_n(n)$ is not a formal theorem of F

is on the face of it not a formal property belonging to F but a meta-property belonging to the metalanguage. It seems implausible that this property has a formal counterpart among the formal properties of F, in particular among the members of the sequence (1).

But Gödel shows that (2) must have such a counterpart in (1)—that a member of the sequence (1) formalizes the metaproperty (2) or, which amounts to the same thing, that this metaproperty is the interpretation or intuitive meaning of a member of the sequence (1). The method by which he shows this is known as the arithmetization (also the 'Gödelization') of the metalanguage or metamathematics, a procedure which is quite analogous to Descartes' arithmetization of Euclidean geometry—the provision of numerical coordinates for non-numerical objects, and of numerical relations for the non-numerical relations between these objects.

To each of the signs of F—*e.g.* $\sim$, $\vee$, (—an integer is assigned so that every finite sequence of signs corresponds to a finite sequence of integers. It is easy to find functions which will establish a one-one correspondence between finite sequences of numbers and numbers. (For example, if we agree to assign to a sequence $n_1, n_2, \ldots, n_m$ the product $p_1{}^{n_1}.p_2{}^{n_2} \ldots p_m{}^{n_m}$, where the p's are the prime numbers in their natural order, it is always possible to reconstruct the sequence from the number by factorization.) In this way every sign, every sequence of signs (*e.g.* every formal proposition) and every sequence of sequences of signs is assigned its numerical coordinate or Gödel number. Statements about formal expressions can thus be replaced by statements about integers.

Again, to every class of expressions there corresponds a class of Gödel numbers. The classes of Gödel numbers needed for the incompleteness theorem are all defined recursively, *i.e.* each element can be actually calculated from the previous ones. The same is true of the required relations between Gödel numbers and of the functions which take Gödel numbers for their arguments and values. It is in particular possible to demarcate in this manner a class T, the class of all formal propositions which are formal theorems in F. (The statement that $p \lor \sim p$ is a formal theorem of F is then equivalently expressed by $c \in T$, where c is the Gödel number of $p \lor \sim p$ in F.) It is equally possible to indicate in this manner a recursive function $\phi(n, p)$ of two integral arguments whose value is the Gödel number of the formal proposition $W_n(p)$, *i.e.* the formal proposition which we get by 'saturating' the nth member of the sequence (1) with the numeral p. After these preparations (which, in the actual proof, naturally take more time, space and effort, and give accordingly more insight into its nature) we can give the Gödel translation of (2), *i.e.* of

$$W_n(n) \text{ is not a formal theorem of } F$$

as

$$(3) \quad \phi(n, n) \text{ non} \in T,$$

i.e. the value of $\phi(n, n)$ is a Gödel number which is not a member of the class T of the Gödel numbers of formal theorems of F.

Now (3) is a property of integers belonging to elementary arithmetic. It must, therefore, have a formalization in F, which must moreover be found in the sequence (1) of $W(.)$'s; for this sequence contains every formal property of numerals. Assume then that we have found that (3) is formalized by the qth member of the sequence, *i.e.* by $W_q(.)$.

The formal property $W_q(.)$ takes numerals as its arguments, among them also the numeral q. We consider therefore the formal proposition $W_q(q)$, which is the undecidable formal proposition we wished to construct. The interpretation of $W_q(q)$ is: the integer q has the property formalized by $W_q(.)$, *i.e.* the arithmetical property: $\phi(n, n)$ non $\in T$; or equivalently: $W_q(q)$ is not a theorem of F.

If $W_q(q)$ were a formal theorem of F it would formalize a false arithmetical proposition. If $\sim W_q(q)$ were a formal theorem of F, then $W_q(q)$ would formalize a true arithmetical proposition. But then a false arithmetical proposition, namely $\sim W_q(q)$, would be formalized by a formal theorem of F. Since *ex hypothesi* F is a consistent formalization of elementary arithmetic, neither case can arise. $W_q(q)$ is undecidable and F is incomplete.

Variants of Gödel's result are obtained by varying the assumptions concerning F, and the methods of proof—all of which, however, allow the actual construction of the desired formal propositions.

The ideas and techniques, especially the arithmetization of metamathematics, which yield the incompleteness theorem and its variants also yield Gödel's second theorem concerning formalisms of type F. If F is consistent and if f is a formalization of the statement that F is consistent, then f is not a formal theorem of F. Briefly, the consistency of F is not provable in F.

The second theorem implies the impossibility of proving the consistency of formalized classical mathematics by finitist methods. For in spite of a certain vagueness in demarcating the notion of finitist proofs, any such proof can be arithmetized and incorporated into F. To prove the consistency of F by finite or 'finitary' means is thus equivalent to proving the consistency of F in F—which by Gödel's second theorem is impossible. The original programme for a consistency proof has to be abandoned, or it has to be relaxed by redefining 'finitist proof'.

We may now make some brief remarks on the theory of recursive functions which was the main instrument of Gödel's proofs. (The remarks follow in the main R. Péter's treatment.) A recursive function is a function which takes non-negative integers as arguments, whose values are again non-negative integers and which is so defined that its values can be 'effectively' calculated. The meaning of 'effective calculation' or 'computability' itself is clarified in developing the theory. The definition of a recursive function does not depend on any assumption either that *there exists* among the totality of integers one which is specified only as having a certain property, or that *all members* of this totality have a certain property. The theory of recursive functions can thus be developed without the universal or existential quantifier. That a large part of arithmetic and logic can be developed in this manner was recognized by Skolem as early as 1923.[1] A main motive for developing this theory was the fact that by abandoning unrestricted quantification, the set-theoretical antinomies can be avoided— 'existence of a set' becoming equivalent with computability of its members.[2]

One of the simplest recursive functions can serve as the definition of adding to a fixed non-negative integer a another integer n. Consider

$$\phi(0, a) = a$$
$$\phi(n+1, a) = \phi(n, a) + 1.$$

The first equation, here, tells us the value of the addition of 0 to a. The second tells us how to find the value of the addition of $n+1$ to a when the value of the addition of n to a has already been found. We can

---

[1] *Begründung der elementaren Arithmetik durch die rekurrierende Denkweise ohne Anwendung scheinbarer Veränderlichen mit unendlichen Ausdehnungsbereich*, Videnskapsselskapets Skrifter 1, Math.—Naturw. Kl. 6, 1923.

[2] See also R. L. Goodstein, *Recursive Number Theory*, Amsterdam, 1957.

thus find the values of the function for $n=0$, $n=1$, $n=2$, $n=3$, etc. They are a, $a+1$, $a+2$, $a+3$, etc. If we write $\beta(a)$ for $a+1$, then $\beta(a)$ expresses the operation of forming the immediate successor of a non-negative integer. Our recursive function can then be written

$$\phi(0, a) = a$$
$$\phi(\beta(n), a) = \beta(\phi(n, a)).$$

In similar fashion we can define multiplication of a fixed positive integer a by a positive integer n. If $\phi(n, a) = n.a$, we have

$$\phi(0, a) = 0$$
$$\phi(n+1, a) = \phi(n, a) + a.$$

In the same way we can define exponentiation and other functions of arithmetic.

The form of these recursive functions is:

$$\phi(0) = K$$
$$\phi(n+1) = \beta(n, \phi(n))$$

Here ϕ is a function of one variable, β a function of two variables, and K a constant or function with no variable. The variable n for which successively 0, 1, 2, etc. are substituted is called the recursion variable. But the values of ϕ and, therefore, β may depend also on other variables which, however, do not enter into the process of recursion, during which they are treated as constants—different values being substituted for them either before or after the recursion, *i.e.* the calculation consisting in the successive substitutions for n. These other variables are, in accordance with the usual terminology of mathematics, called 'parameters'. A definition of the form

$$\phi(0, a_1, a_2, \ldots, a_r) = \alpha(a_1, a_2, \ldots, a_r)$$
$$\phi(n+1, a_1, \ldots, a_r) = \beta(n, a_1, a_2, \ldots, a_r, \phi(n_1, a_1, a_2, \ldots, a_r))$$

is called a *primitive recursion*.

If two functions are given we may form a new function by substituting one function for one variable in the other, *e.g.* from $\phi(x, y, z)$ and $\psi(u)$ we can get by substitution $\phi(\psi(u), y, z)$, $\phi(x, y, \psi(u))$, $\psi(\phi(x, y, z))$, etc. Primitive recursions and substitutions yield a large and important class of functions called *primitive recursive functions* characterized [1] as those functions whose arguments and values are non-negative integers and which starting from 0 and $n+1$ are defined by a finite number of substitutions and primitive recursions.

[1] Péter, *op. cit.*, p. 32.

In his proofs Gödel used only primitive recursive functions. To see how formal properties can be arithmetized we consider the definition of recursive relations. A relation $B(a_1, \ldots, a_r)$ is primitive recursive if there exists a primitive recursive function $\beta(a_1, \ldots, a_r)$, such that it equals 0, if and only if the relation B holds between $a_1, \ldots, a_r$. If $W(a)$ is a property, it is primitive recursive provided there exists a primitive recursive function which equals 0, if and only if a has W. The complementary relation $B'(a_1, \ldots, a_r)$ of $B(a_1, \ldots, a_r)$ is also primitive recursive and holds only if $\beta(a_1, \ldots, a_r) \neq 0$. In this way the notions 'being a complement', 'being a conjunction' and more complex notions of methamathematics including 'being a formal theorem of F' become expressible as primitive recursive functions, and relations between Gödel numbers.

It follows from a theorem of Turing (1937) that the computation of any primitive recursive function can be left to a machine. In fact he showed that a wider class of functions, the so-called *general recursive functions*, are computable by Turing-machines. Before this was shown, Church had proposed that the rather vague notion of effective computability should be analysed as solvability by general recursive functions. This proposal was justified by Church's own results and by other results which, though at first sight unconnected, all proved equivalent. As regards this problem of identifying effective computability with solvability by general recursive functions, expert opinion is no longer undivided.[1] On this question nothing can profitably be said in the present context by the present author. The theory is developing into a new branch of pure mathematics whose relevance to the problems raised by Hilbert is merely one of its important aspects, and perhaps no longer the most important.[2]

[1] See Péter, *op. cit.*, §§ 20–22.
[2] See Myhill, *op. cit.*, p. 136.

V

MATHEMATICS AS THE SCIENCE OF
FORMAL SYSTEMS: CRITICISM

IN discussing the formalist philosophy of mathematics we shall continue our plan of examining its account of pure mathematics, of applied mathematics and of the notion of infinity. As in the case of logicism, we shall consider first of all the simple examples of '$1+1=2$' and of 'one apple and one apple make two apples'.

The formalist distinguishes, as we have seen, between the sequence of marks $\langle 1+1=2 \rangle$ (the formula) and the true statement *that* this formula or the process by which it is produced has certain literally formal characteristics—the characteristics, as it has been put, of being a theorem-formula. The sequence $\langle 1+1=2 \rangle$ is not a statement but a physical object, and so is neither true nor false. What is true-or-false, is the statement that this sequence $\langle 1+1=2 \rangle$ is a theorem-formula. In other words whereas on the logicist view the certainty of the mathematical statement has its root in logic, its certainty on the formalist view springs from the indubitableness of the description it gives of very simple experimental or physical situations. For the logicist arithmetical statements are disguised logical statements. For the formalist they are disguised empirical statements. And as it was necessary for us to examine, and in the end to reject, the former claim, so we must examine the latter also.

The formalist claim is surrounded from the outset by an air of paradox. The impression seems to come from two sources: on the one hand his apparent assumption that only three positions are possible, namely (a) the statements of pure mathematics are logical, (b) they are synthetic *a priori* in Kant's sense and (c) they are empirical; on the other hand it comes from his apparent conviction that the first possibility has been demonstrated not to hold and that the second must be ruled out as too obscure and as inadequate to the variety of different mathematical systems. But however modestly, however silently, the claim that the statements of pure mathematics are empirical is being made.

Turning now to the statements of applied mathematics—about the physical addition of apples, etc.—we find the position of the formalist again calling for comment. One might perhaps at first sight feel tempted to say, here, that a straightforward one-one correspondence does indeed hold between the (metamathematical) statement that $\langle 1+1=2\rangle$ is a theorem-formula and the statement that one apple and one apple make two apples. All that *seems* to be required is to put apples and apple-operations in the place of strokes and stroke-operations. Yet the situation is not quite so simple.

It is instructive to compare the problem of applied mathematics as it faces the formalist and as it faces the logicist. The logicist has to consider '$1+1=2$' in the logical transcription

(1) $(x)(y)(((x \in 1) \,\&\, (y \in 1)) \equiv ((x \cup y) \in 2))$ (see p. 53)[1]

And he has further to consider two versions of 'one apple and one apple make two apples', namely

(2a) $((a \in 1) \,\&\, (b \in 1)) \equiv ((a \cup b) \in 2)$

where a and b are two specified unit classes with no common elements and where $(a \cup b)$ is their logical sum, and

(2b) an empirical law of nature about the behaviour of apples.

As we have argued, if (1) is logical then (2a) is also logical. But unlike (2a), (2b) is empirical; and the logicist owes us some explanation of the relation between the *logical* statement (1) and the *empirical* statement (2b). In criticizing logicism we have argued that the concepts—the number-concepts and the concept of addition—are different in (1) and in (2b); and that, by not discerning this difference, but, on the contrary, conflating non-empirical and empirical concepts, logicism fails even to state the problem of the relation between (1) and (2b), to say nothing of proposing a solution.

The situation which faces the formalist is somewhat similar. He has to consider (1) the mathematical statement '$1+1=2$' in metamathematical interpretation, namely: 'On the one hand putting 1 after 1; and on the other hand producing the sequence 11 (both operations performed in accordance with the rules for juxtaposition) lead to the same stroke-expression, namely 11.' We shall say that this and similar metamathematical statements, which are capable of demonstration *ad oculos*, are empirically evident. The formalist has further to consider two interpretations of 'one apple and one apple make two apples' namely:

(3a) a statement which differs from the statement about the juxtaposition of strokes only in being about apples; and

[1] x and y are assumed to be distinct and not empty.

(3b) a statement about some physical addition of apples (putting them in drawers and taking them out after a few days, etc.) which is *not* defined in accordance with the rules governing juxtaposition, as considered in metamathematics.

The point is this: if the statements of pure mathematics are empirically evident then (1) and (3a) are empirically evident. But (3b) and the mathematically-formulated laws of nature, are not—or not in the same sense of the terms—empirically evident. Thus, just as, on the one hand, the logicist is faced with the problem of explaining the relationship between the allegedly logical (and, we may say, logically evident) statements of pure mathematics and empirical statements of applied mathematics, so on the other hand the formalist is faced with the problem of explaining the relationship between empirically evident statements of pure mathematics and empirically not evident statements of applied mathematics.

Apart from the general problem of the logical status and function of the propositions of pure and of applied mathematics just mentioned, every philosophy of mathematics must also hold a position in respect to the question of infinite totalities. The formalist, as we have seen, does not allow himself the assumption of actually infinite sets, or the use of transfinite methods, within metamathematics. He does, however, allow the use of *symbols* for actually infinite entities. These symbols he regards as perceptual objects after the fashion of stroke-expressions, within the regulated activity of manipulating marks, which is the perceptual subject matter of metamathematics. We have seen that Hilbert's programme of establishing a one-one correspondence between all (apparently innocuous) propositions of classical mathematics on the one hand and formulae embodying them has been shown by Gödel to be impossible, or at least highly problematic. This raises, among others, the question as to whether or in what sense the adjunction of formulae embodying statements about infinite totalities to the formulae of elementary arithmetic can be regarded as justifying the use of actual infinities in classical mathematics. It is a question again of 'Cantor's paradise'. Has Gödel's proof of incompleteness expelled us from it or only made its territory smaller?

Connected with all these problems, in their formalist variants, is the relationship between logic and mathematics. The formalist's metamathematical reasonings appear as something read off from indubitable perceptual experience, as *demonstrationes ad oculos*, neither in need of justification nor capable of it. We must, therefore, consider the formalist position regarding the nature of this intuitive logic, its demarcation from a non-intuitive logic, the relation of each to the other and of both to mathematics.

A reasonable plan of discussion would seem to be, first, to consider the formalist's conception of pure mathematics as the manipulation or construction of strings of symbols and of pure metamathematics as consisting of (i) statements which are empirically evident and (ii) of reasonings involving only such statements. Next would have to be considered the formalist's account of the relationship between these empirically evident propositions of pure mathematics and the propositions of applied mathematics, which are not capable of demonstration *ad oculos*, but depend for their verification on the ordinary experimental and observational techniques of natural science. The problem of infinity could then be discussed as part of the general problem of the nature of the alleged adjunction of ideal to real entities in classical mathematics, the corresponding adjunction of symbols for ideal to symbols for real entities, and the metamathematical statements about the handling of both types of symbol in a purely formal manner; our discussion being concluded by an examination of the formalist conception of logic.

1. *The formalist account of pure mathematics*

According to the formalist account, metamathematical statements either describe certain types of symbol-manipulation or express inferential relations between such descriptive statements. Thus to all appearance the statements of metamathematics are empirical statements, the concepts applied in making them are empirical concepts, and the inferences concerned are always from empirical statements or concepts to empirical statements or concepts.

It is, of course, quite clear that the statements of metamathematics are not about operations on particular strokes. The strokes are replaceable by others without changing the content of the statements. The strokes, the stones, shells, etc., on which the operations are actually performed are merely representative—tokens of types, to use a distinction made by C. S. Peirce. In describing a particular operation, say the elementary addition of 1 to 1, by saying that it yields the figure 11, what matters is the relation between the two concepts, 'elementary addition' and 'figure of type 11'; and this relation is unaffected by substituting other tokens for these particular strokes. The concepts (predicates, attributes, etc.), however, are instantiated by perceptual objects and operations; and it is this easily-available or easily-constructible perceptual situation which allows the relation between the concepts to be 'read off' from it. This perceptual situation we can bring about (almost) whenever we like, and this perceptual situation it is, which allows us to 'demonstrate *ad oculos*' the relation between the concepts.

However, the formalist's concept 'x is a stroke' does not strictly speaking describe or, what comes to the same thing, is not strictly speaking instantiated, or exemplified by, any particular stroke. Thus the formalist cannot but assume that the strokes with which he is dealing have certain properties which those found in perception do not have. No physical or perceptual stroke is permanent although in making it the object of metamathematics, we quite properly regard it as such. We 'abstract' in one way or another from its impermanence.

The point may seem a trivial one. Yet, on the one hand it is considered of sufficient weight by Frege[1] to serve as a chief ground of his attack upon earlier variants of formalism, variants which in this respect do not differ from its most modern forms; while on the other hand contemporary formalists and those philosophically near them also see the need for justification, if they are to identify physical or perceptual strokes, which are impermanent, with the instances of the concept 'x is a stroke' in the sense in which it entails 'x is permanent'. It is instructive to quote, here, a relevant passage from chapter I of an important recent work[2]: 'Let it be said, to prevent misunderstanding that the subject of these investigations is not the individual realizations of the figures. Thus if, say, the figures 1, 11, 111, . . . which are composed only from 1, are called "numbers", this does not imply that when the realizations now seen by the reader have perhaps rotted away there will be no numbers. Anybody who has the capacity of producing such figures can at any time meaningfully speak of numbers.'

One might put the case by saying that the stroke-expressions are treated *as if* they were permanent. But this means that the stroke-symbols which are written down on paper are not the subject-matter of metamathematics. It means that 'x is a permanent stroke-symbol' rather than 'x is an (impermanent) stroke-symbol' is a metamathematical concept. It is indeed a matter of small importance if, in the cause of calculation or of metamathematical reasoning, one ignores the difference between the two concepts. The difference is, however, important for the philosophy of mathematics. It implies that it is

---

1 *Grundgesetze*, vol. 2, §§ 86 ff.; also translations by Geach and Black pp. 82 ff.
2 P. Lorenzen, *Einführung in die operative Logik und Mathematik*, Berlin, 1955.

strictly speaking false that the categorical statements of meta-mathematics are perceptual statements of indubitable certainty. They may be of indubitable certainty, but they are not perceptual. Strokes on paper are instantiations of 'x is an impermanent stroke'.

Another stipulation which the formalist is bound to make concerns the definiteness or exactness of those stroke-expressions and operations, which form the subject-matter of his science. He stipulates for example that every stroke-expression, *i.e.* every instance of 'x is like 1' or every instance of 'x is like 11', is perfectly definite or, what comes to the same thing, that the corresponding concepts are exact in the sense of not admitting of border-line cases. Yet the two concepts do not only admit of separate, but also of joint border-line cases. If we bring 1 and 1 more and more closely together we reach a stroke-figure of which it is equally correct to say that it is like 1 and also that it is like 11. While 'x is like 1' and 'x is like 11' admit of common border-line cases, the corresponding metamathematical concepts do not admit of these: they are exact concepts.

One might again say that the strokes which are the formalist's concern can be treated *as if* they were instances of an exact concept 'x is a stroke'. But this does not mean that the concept 'x is a (perceptual) stroke', which admits of definite instances, does not *also* admit of border-line cases. What has been said about the difference between the perceptually instantiated 'x is a (perceptual) stroke' and the metamathematical 'x is a (permanent) stroke' applies also to the difference between the inexactness of the perceptual and the exactness of the metamathematical concept.

Let me say again that the difference between a concept which is instantiated in perception and the corresponding metamathematical concept which is, roughly speaking, an idealization of the former is not important to the working mathematician, including the working metamathematician. That it is important to the philosophy of mathematics was not only recognized by Frege, but also by Plato, Leibniz, and Kant. Plato distinguishes sharply between diagrams on the one hand and the Forms and their instances on the other. According to him a diagram, such as a perceivable stroke on paper, is not an instance of the Form of the number 1; it only 'tries to be like it' or approximates to it. Between the perceptual stroke and the number, the relation is not instantiation but 'participation' or μέθεξις. Neither does Leibniz identify the figures of his *characteristica univer-salis*, which in so far as they are used in mathematics express timeless 'truths of reason', with these truths or the universals which are the constituents of them. Similarly Kant carefully distinguishes between statements about physical or perceptual objects, such as marks on paper, and statements about constructions in the space of pure perception or intuition—a distinction which Brouwer and his intuitionist followers endorse.

We find thus, just as we did previously in discussing the logicist philosophy of mathematics, a conflation of two different number-concepts. In the case of logicism, inexact empirical concepts which can be instantiated in perception are conflated with exact concepts which cannot: 'Natural Number' and 'natural number' are not kept separate. The logicist emphasis is on the exact and *a priori* concepts which according to him are translatable into concepts of logic. In the case of formalism the emphasis is on an allegedly empirical concept of number which is regarded as both exact and descriptive of perceptual data. The fact that the exactness of the formalist number-concept is due to an idealization is overlooked or ignored.

There is no need to repeat what has been said earlier about the difference between the two types of concept especially as the subject will come up for a fuller discussion later, independently of the exposition and criticism of either logicism or formalism. However, if the distinction is justified for number-concepts, it has to be granted also for the concept of operations on numbers. Indeed, just as we had to distinguish, *e.g.*, between the mathematical operation of forming the logical sum of (non-empirical) natural numbers and the physical operation of adding (empirical) Natural Numbers, so we must distinguish between the mathematical operation of adding idealized strokes and the physical operation of physically adding physical strokes —or apples or stones or other such perceptually given objects.

Since the concepts of metamathematics and the statements in which these concepts are applied are not empirical, their subject-matter is also not empirical. The strokes on paper and the operations upon them are just as little the subject-matter of metamathematics as figures and constructions on paper are the subject-matter of Euclidean geometry. Both types of marks and constructions are diagrammatic; and diagrams, however useful and practically indispensable, are 'representations' which are neither identical nor isomorphic with what they are used to 'represent'. A diagram is in this respect like a map which 'represents' a country or like Ariadne's thread which led through the labyrinth. It is not like the country or the labyrinth itself. (I have put 'represent' in quotes in order to indicate that I am not using the term in its (now perhaps) dominant sense implying isomorphism between representing and represented system. The inexactness of the empirical and the exactness of the non-empirical concepts precludes isomorphism between the instances and relations of the two systems.)

If we accept the preceding distinctions between empirical concepts and operations and the non-empirical ones then a science of empirical stroke-expressions and operations must be distinguished from the science of idealized stroke-expressions and operations. The latter alone would—if the qualifications concerning the permanence and definiteness of the stroke-expressions are taken seriously—be meta-mathematics. The former science might for the moment be called by the barbarous name 'diagrammatics' to imply that it is concerned with perceptual diagrams. Diagrammatics serves metamathematics in a way which is analogous to the service rendered by cartography to geographical exploration. (The analogy, of course, breaks down when we consider that the subject of diagrammatics is perceptual and the subject of metamathematics non-perceptual entities, whereas the subject of cartography and geographical exploration are both perceptual.)

We may even develop the analogy a little. Unless countries existed there could be no maps, since maps, even those which do not map any country, are defined in terms of the relationship of some drawings to some countries. Similarly unless mathematical systems existed there could be no diagrammatic formalism, since even those which do not formalize any mathematical theory are defined in terms of the relationship of some formalisms to some theories. The difference between maps and non-maps is thus analogous to the difference between diagrammatic formalisms and non-formalisms and the difference between mapping and non-mapping maps to the difference between formalizing and non-formalizing formalisms. Not every coloured piece of paper is a map, not every map maps a country. Not every game with strokes is a diagrammatic formalism, not every diagrammatic formalism is a formalizing one. Curry's demarcation of the notion of a formalism (apart from any interpretation) corresponds to the cartographical demarcation of the notion of a map (apart from any interpretation).

The mathematician needs diagrammatic figures as instruments, just as the geographical explorer needs maps. Without some knowledge of diagrammatics the mathematician cannot do his job of thinking within a mathematical theory or of devising a new one. Similarly without some knowledge of cartography an explorer cannot explore a country which has been discovered or which he discovers for the first time. Cartography and diagrammatics are thus auxiliary sciences for geographical exploration and mathematics respectively. Analogy, especially in this case, which has only very limited scope, cannot take the place of argument. It does, however, emphasize that the distinction between empirical and non-empirical number-concepts in no way implies that notation, symbolism and the science of diagrammatic formalisms are not of the highest importance heuristically and in many other respects. Their great importance has never been denied by the philosophers who distinguish between statements about perceptual and statements about mathematical objects. Thus, according to Plato, the mathematician although inquiring into relationships between eternal Forms must for practical reasons use physical diagrams to aid his inquiry. Modern intuitionists, who deny that mathematics is concerned with rule-governed manipulations of marks on paper or other perceptual objects, are similarly convinced of the usefulness and practical necessity of (diagrammatic) formalisms and have constructed a number of such in order to convey their mathematical ideas.

Formalist philosophers of mathematics are, as has been pointed out earlier, very clearly aware of the idealized, non-empirical character of certain mathematical concepts, in particular the notion of actual infinities. Yet they do not seem to realize that the disconnection of mathematics from perception—the introduction of non-empirical concepts—takes place not at the transition from the elementary arithmetic to the amplified, but at the very outset of any development of the elementary. The concept of natural number itself is non-empirical.

2. *The formalist account of applied mathematics*

For the logicist the problem of applied mathematics consists in relating allegedly logical statements, which cannot possibly be false, to empirical statements which may be false. For the formalist the problem consists in relating empirically evident perceptual statements, which cannot be false, to empirical statements which may be. We consider, as before, the metamathematical statement '$1 + 1 = 2$' which describes the juxtaposition of two strokes and its result and the statement of a very simple physics 'one apple and one apple make two apples' (*i.e.* case 3b on p. 100).

If we regard the apple-statement as being the result of replacing in the situation described by '$1 + 1 = 2$' each stroke by an apple and the juxaposition of strokes by the juxtaposition of apples, then either both statements will be self-evidently true or neither will be. This is, however, incompatible with the assumption with which we started. If the stroke-statement is self-evidently true and the apple-statement is not, there must be a difference between the metamathematician's strokes and stroke-juxtaposition on the one hand and apples and apple-juxtaposition on the other. The difference is, of course, that the strokes, and the stroke-juxtaposition, are instances of non-empirical, idealized, exact concepts; whereas the apples and the apple-juxtaposition are instances of empirical, inexact concepts. Just as we had to distinguish between 'addition' as a logical sum and 'addition' as a physical operation, so we must distinguish between 'addition' as mathematical juxtaposition and 'addition' as a physical operation.

Replacement of the kind described cannot lead from the meta-mathematical to the physical statement. Perhaps because this is quite clearly not the case, formalists do not suggest that statements of applied mathematics—in particular of applied elementary arithmetic—are replacement instances of stroke-statements. In order to preserve the self-evident character of the stroke-statements and the character other than self-evident (conjectural, inductive, probable, corrigible, etc.) which belongs to the apple-statements, a more complex analysis is required.

The relationship is often explained in terms of one-one corres-pondences [1] between strokes and stroke-relations on the one hand and, e.g., apples and apple-relations on the other. But such a one-one corres-pondence cannot be established. The metamathematical concept 'x is one stroke' (either *ex hypothesi*, or by the decree of the formalist philosophers) does not admit of border-line cases. It is never correct to say of an object 'This is one stroke' unless it is incorrect to say of it 'This is not one stroke' but, for example, 'This is two strokes'. The case is, of course, quite different with 'x is one apple'. The rules governing this expression do indeed permit us to say sometimes with equal correctness, of a fruit growing on an apple-tree, that it is one apple and also that it is two apples grown together. The fruit in question does not stand in a one-one relation to either an instance of 'x is a one-stroke' or to an instance of 'x is not a one-stroke'. It is a border-line or neutral case of an inexact concept which admits of positive, negative and neutral cases; whereas the concept 'x is one stroke' admits of positive and negative cases only. We may, of course, decide to deal with 'x is one apple' *as if* it had no borderline cases. But then we should idealize the concept in the same manner as the metamathematician idealizes the concept of a physical stroke. We should then again compare idealized strokes with idealized apples, whereas of course the point of bringing in the notion of one-one correspondence was to relate ideal strokes to real apples—and meta-mathematical statements about the former with physical statements about the latter.

---

[1] *Zuordnungsdefinitionen*, 'co-ordinative' or 'correlating' definitions, *e.g.* Reichenbach, *Wahrscheinlichkeitslehre*, Leiden, 1935, pp. 48 ff.

108   THE PHILOSOPHY OF MATHEMATICS

The impossibility of establishing the required one-one correspondence of the instances of 'x is an (ideal) stroke' or any other exact
concept, with the instances of 'x is an apple' or any other inexact
concept, becomes even more clear if we compare the situation with a
purely arithmetical one. Let us call an integer a P-number (to remind
us of positive cases) if it is divisible by 2 but not by 3; an N-number
(to remind us of negative cases) if it is divisible by 3 but not by 2; and
a B-number (to remind us of borderline cases) if it is divisible by 2
and by 3. A P-N-set, which consists of P- and N-numbers, thus corresponds to an exact concept; a P-N-B-set, consisting of P-, N- and
B-numbers, would seem to correspond to an inexact concept. It is
trivially clear that we cannot establish a one-one correspondence
between *any* P-N-B-set and *any* P-N-set with the P-numbers of the
first set standing in one-one correspondence with the P-numbers of
the second, the N-numbers of the first standing in one-one correspondence to the N-numbers of the second, and no numbers of the
first set being left over. So far the comparison between P-N-B- and
P-N-sets is analogous to that between inexact and exact concepts.

Yet the second case is more complex as is also more or less
obvious. The reason is simple. The B-cases of a P-N-B-set are sharply
defined—it is quite clear of any given integer whether it is divisible by
2 and 3. No such clear demarcation is possible for the borderline cases
of 'x is one apple' and any inexact concept. We might of course *try*
to collect all the borderline cases between 'x is one apple' and 'x is a
couple of apples'—calling the concept under which they fall, say,
'x is an apple-twin'. But this concept—unlike the set of B-numbers—
would again admit of borderline cases, both between cases of 'x is one
apple' and 'x is an apple-twin' and between 'x is an apple-twin'
and 'x is a couple of apples', and the collection of these borderline
cases would again lead to new borderline cases.

What has been said about the relation between ideal strokes and
physical apples applies, and for similar reasons, also to the relation
between mathematical addition—an ideal juxtaposition of such
strokes—and the various different operations of physical addition.
'x is the result of a juxtaposition of ideal strokes' does not admit of
borderline cases whereas 'x is the result of a physical addition' does.
The contrast between mathematical addition, as conceived by the
formalist, and all physical additions is quite as fundamental as the
contrast between mathematical addition in its logicist conception as a
logical sum and physical additions.

In formalist philosophies of mathematics, in particular that of
Hilbert and Bernays, recognition is given, as we have seen, to the

importance of distinguishing between concepts which are and such as are not instantiated in perception. But these philosophers draw the line at a different place. For them infinite totalities of stroke-expressions cannot possibly be instantiated in perception and it would be pointless to look for an infinite totality, say, of apples. I have argued in favour of drawing the line much earlier. Even the notions of elementary arithmetic—apart from any adjunction to them of infinite totalities—are not instantiated in perception. It is consequently point-less to look for a one-one correspondence even between instances of 'x is one stroke' (in the sense defined by the formalist) and the instances of 'x is one apple'.

Once we decide that there can be no one-one correspondence between the entities and operations of elementary arithmetic, which are instances of exact concepts, and perceptual entities and operations, which are instances of inexact concepts, an altogether different account of applied mathematics seems called for. We may on the one hand deny that there is a sharp difference between the allegedly indubitable state-ments of metamathematics and the ordinary empirical statements of applied mathematics. One might say instead that the difference is merely pragmatic—consisting in our greater reluctance to reject the metamathematical statements from the body of our beliefs than to reject the mathematically expressed laws of nature. This type of formalist pragmatism would be analogous to the logicist pragmatism which we discussed in chapter III. Just as logicist pragmatism admits only a difference of degree between allegedly logical statements and empirical statements, so formalist pragmatism would only admit a difference of degree between empirically evident metamathematical and other empirical statements. But this kind of formalist pragmatism would conflict with the basic theses of the original formalism, just as, *e.g.*, Quine's pragmatism conflicts with the basic theses of Frege and Russell.

The reason for this is clear. Just as the original logicism assumes that there is a sharp distinction between logical and empirical propo-sitions to which every execution of the logicist programme must conform, so the original formalism assumed a sharp distinction between metamathematical and extra-mathematical statements even if both are empirical. Whether one decides to call the logicist prag-matists 'logicists' at all, or the formalist pragmatists of the kind described 'formalists', is of course largely a verbal question. It would seem on the whole that philosophers who deny any but differences of degree between logical and non-logical propositions are still called 'logicists'—the reason being that some of those who have made important contributions to the logicist cause have at the same time, or later, embraced pragmatism. Those who deny any but differences in degree between metamathematical and other empirical propositions are not normally called formalists. The reason again seems to be that those who have contributed to the formalist cause—*e.g.* Hilbert, Bernays and Curry—have not at the same time also been pragmatist in the radical sense in which the term is usually employed.

We have seen that attempts to explain the relation between '1 + 1 = 2' and 'one apple and one apple make two apples' by regarding the latter statement as either (a) a replacement-instance of the former ('apple' for 'stroke', 'physical addition' for 'stroke-juxtaposition') or (b) as being isomorphic with the former end in ignoring the very difference between the two statements which they presuppose and indeed which they set out to explain. A much more promising attempt is, as we have already seen, due to Curry.

Curry distinguishes between on the one hand the *truth* of (metamathematical) statements about marks and strings of marks, *i.e.* the truth of the statements of pure mathematics, and on the other the *acceptability* of a theory of pure mathematics for a given purpose. The truth of the statements of pure mathematics is according to him 'an objective matter about which we can all agree' whereas the applicability of a theory of pure mathematics 'may involve extraneous considerations'.[1] He illustrates his distinction by comparing a formalized system of classical analysis with its applicability in physics. The fact that physicists 'associate' with the predicates of classical analysis certain physical notions, has proved extremely useful. It is a pragmatically justifiable procedure. A formalized system of classical analysis, even if it is demonstrably inconsistent, is to be preferred for the purposes of physics to a formalized system fulfilling the finitist requirements of intuitionist philosophers. The point is made emphatically in the following passage: 'The intuitionistic theories are so complicated that they are utterly useless; whereas the classical analysis has been extremely fruitful. This is the decisive point; and so long as this usefulness persists, classical analysis needs no other justification whatever.'[2]

Curry thus insists, as I believe rightly, on a sharp distinction between pure and applied mathematics. His pragmatism extends to applied mathematics only and does not imply the radical pragmatist thesis that the statements of pure and of applied mathematics do not differ in kind but only in degree according to the greater or smaller reluctance with which we are prepared to drop them.

[1] *Op. cit.*, p. 60.   [2] *Op. cit.*, p. 61.

Yet, if we may so put it, his pragmatism with respect to applied mathematics is rather coarsegrained. The reason for this lies in his failure to analyse the relation between the formal predicates of pure mathematics and the empirical notions of applied mathematics. He says no more than that they are 'associated' and by implication that they are not isomorphic. Indeed, if one were to undertake a finer analysis of the way in which the formal and the empirical notions are associated one would have to supplement Curry's thorough analysis of formal predicates by a similarly thorough analysis of empirical predicates. This he does not provide. Nor does he, on the other hand, argue that it is impossible.

3. *The concept of actual infinity*

Towards the employment of the concept of an actual infinity three philosophical attitudes have in the main been taken up which may be called respectively finitism, transfinitism and methodological transfinitism. Finitists such as Aristotle, Gauss, and the older and the new intuitionists, deny all 'real' content or even 'intelligibility' to such mathematical notions as are not characteristic either of finite aggregates or at most of potentially infinite, *i.e.* growing but never completed, aggregates. (Those among them who do not admit even the conception of potentially infinite aggregates might be called 'strict finitists'.) Transfinitists such as Cantor and his followers ascribe the same reality and intelligibility to transfinite as to finite concepts. Methodological transfinitists, in particular Hilbert, admit transfinite concepts into mathematical theories without according them full 'ontological' status. They are admitted because they are useful for such purposes as the simplification and unification of mathematical theories.

Each of these philosophical doctrines will remain a lifeless dogma of an autonomous and selfsufficient metaphysics, unless it also functions as a regulative or directive principle, a programme which is to be satisfied in the construction of mathematical theories. The distinction between dogma and programme is important for an understanding of the nature of the controversies concerning the notion of actual infinity and indeed of many other philosophical controversies.

A dogma is a proposition which, if meaningful at all, is either true or false, and of two incompatible dogmas one at least must be false. Thus, *e.g.*, finitism or transfinitism must be false if they are statements about the nature of reality. Yet it is difficult to see how one could decide in favour of either one of them except by embracing it as one might a religious faith.

A programme is quite different. It is not either true or false. If two programmes are incompatible it does *not* follow that one of them is false. But a programme is satisfiable or not-satisfiable and a person who adopts one does (usually) believe it to be satisfiable. To overlook the difference between the *satisfiability* of a programme and the *truth* of a true-or-false proposition is a confusion. Of two incompatible true-or-false propositions one, at least, must be false. But of two incompatible programmes both may be satisfiable.

The division into finitists, transfinitists and methodological transfinitists reminds one of a more general division in philosophy—that between positivists, metaphysical realists and methodological realists. Positivists accord full 'reality' and 'intelligibility' to empirical concepts only, metaphysical realists accord these honours also to some non-empirical concepts, and lastly methodological realists admit certain non-empirical concepts in a purely auxiliary capacity. The positivist programme is satisfied, for example, by phenomenological physical theories such as phenomenological thermodynamics; that of metaphysical realism by Boltzmann's conception of the kinetic theory of gases; and that of methodological realism by theories which admit unobservable entities and their characteristics only with due reserve and qualifications. Needless to add, the notion of what constitutes an empirical concept is usually not sharply defined and varies confusingly from one group of thinkers to another.

These distinctions enable us to be brief in formulating Hilbert's philosophy. He is best understood as a methodological transfinitist who is also a methodological realist. He accords 'reality' only to empirical concepts and holds that such non-empirical concepts as 'actual infinity' should only be admitted if the theory which employs them can be proved to be consistent.

Now I have argued earlier that empirical concepts such as the characteristics of physical strokes and stroke-operations are inexact and that the concepts of arithmetic—including elementary arithmetic —are exact and non-empirical. The line, therefore, which divides finite arithmetic from transfinite is not the same line as divides empirical from non-empirical concepts. Finite and transfinite arithmetic both lie on the non-empirical side. This means further that the positivist metaphysician is also wrong in arguing against transfinite arithmetic on the ground that it operates with non-empirical concepts. In fact to exclude non-empirical concepts from mathematics is precisely to exclude from mathematics the mathematical concepts, *i.e.* to extinguish the subject.

There can be no doubt that the difficulties in constructing a consistent mathematical theory increase as one moves from elementary arithmetic to an arithmetic involving the concept of the totality of all integers; and these difficulties doubtless still further increase when one admits aggregates of cardinal number higher than a. It might further be allowed that each of these steps implies a higher degree of idealization—a further remove from perception. But these, in the present context, are not the significant considerations. So long as the programme of methodological transfinitism was not shown to be unsatisfiable, it might persist. Arguments to the effect that it admits idealizations into mathematics carry no conviction, and arguments that it admits too radical idealizations carry very little. If, in historical fact, a programme has remained unfulfilled, in spite of the greatest efforts by the most competent thinkers over a prolonged period, it is that fact rather than any other which—in politics, science, mathematics and in other fields of human endeavour—causes people ultimately to abandon or modify the programme. This is why metaphysical programmes and the theses with which they are associated, although unrefuted, do not die suddenly but rather fade away.

What an opponent of the formalist programme had to do, if he were to be immediately effective, was to prove that a programme of admitting transfinite concepts into mathematical theories could not, in the nature of the case, be satisfied. Now in its original form this programme was first so to formalize elementary arithmetic and a sufficient portion of the transfinite arithmetic that the formal consistency of the formalism would correspond to the logical consistency of the formalized theory and secondly to prove *by finite methods* the (formal) consistency of the formalism. This programme demonstrably cannot be satisfied since, as Gödel has proved, no formalism of the type here used can formalize arithmetic—even elementary arithmetic—completely.

Gödel's results were published in the period between the composition of volume I of the classic work by Hilbert and Bernays and the appearance of volume II; and that their importance was clearly recognized by the authors themselves is obvious from their preface to the second volume, one of whose (two) central themes is the situation which made it necessary 'to widen the frame-work of the concrete (*inhaltliche*) methods of inference admitted for the theory of proof in opposition to the previous demarcation of the "finite point of view"'.[1] Transfinite induction, which proceeds not through the sequence of natural numbers, but through 'larger', well-ordered sets, is admitted.[2]

[1] *Op. cit.*, vol. 2, p. vii.
[2] See, *e.g.*, chapter V of R. L. Wilder's *Foundations of Mathematics*, New York, 1952.

The question which arises here is whether the admission of transfinite methods of reasoning into metamathematics may not mean abandoning the position of *methodological* transfinitism, the view which allows transfinite concepts *within* mathematical theories and within their formalizations but does not allow them in the non-formalized statements *about* formalisms. The situation now facing the formalist, concerning the distinction between admissible and non-admissible transfinite methods, is quite similar to the situation which, after the discovery of antinomies in Frege's system, faced the logicist.

Logicism started from the assumption that logical concepts and propositions could be clearly distinguished from non-logical. In the course of trying to fulfil the logicist programme of deducing mathematics from logic, the original distinction, which was never too clear, had to be blurred by introducing non-logical, or at least not-obviously-logical, propositions into the premises from which mathematics was deduced. In a similar way formalism made the initial assumption that there was a clear distinction between finite concepts, propositions and finite (*ad oculos*) demonstrations and, on the other hand, transfinite ones. In the course of trying to fulfil the programme of proving the consistency of formalized classical mathematics by finite methods it became necessary to admit transfinite methods too.

In the case of formalism, the attempt to fulfil the original programme fails not only in (what is the historical fact) that the programme has not been satisfied. The failure is that it has turned out to be unsatisfiable. It is true that many results of varying importance have been collected on the way. But as regards the specific philosophical theses and programmes of the original formalism—these have had to be supplemented by *ad hoc* qualifications.

In conclusion we must ask what light the formalist philosophy of mathematics has thrown on the whole notion of actual infinity. Hilbert, we have seen, regarded this notion as a Kantian Idea, a notion which is neither abstracted from perception nor applicable to it and which yet can be introduced into theories without inconsistency. He set out to give both a more precise analysis of this Idea and a rigorous proof of its innocuousness in formalizable systems of classical analysis. On the one hand the Idea of an actual infinity and the propositions involving it are, according to Hilbert, exactly like finite mathematical concepts and statements, and like statements involving them, in being capable of incorporation—without the meaning which they may or may not have—into a complete and consistent formalism. On the other hand the Idea and the propositions involving it are, *unlike* finite mathematical concepts and propositions, incapable of being interpreted as characterizing perceptual characteristics of (very simple) perceptual data. In the case of the objects of the formalism which embody finite notions and propositions, the rules of manipulating these objects *qua* objects can be supplemented by other rules which, together with the former, govern the use of the objects as perceptual characteristics and perceptual propositions. In the case of objects which do not embody finite concepts or propositions this cannot be done.

The refinement which Hilbert attempted of the Kantian account of the Ideas, in particular of actual infinity, consists in bringing in the notions of complete formalization and of formal consistency proof. Owing to the impossibility of proving the consistency of a completely formalized arithmetic, the refinement must be judged unsuccessful. Hilbert's account of the notion of actual infinity is indeed superior to the logicist's employment of it without clarification. His attempt at a consistency proof is, perhaps, superior to Kant's in that it is amenable to more definite procedures. Its failure suggests a modification of the original programme and is a source of much fruitful mathematics. But the logical status of the notion of an actual infinity, as opposed to the readiness of some and the hesitation of others to accord it full metaphysical honours, is still left in the dark—or it is left in the twilight into which Kant had moved it from obscurity.

4. *The formalist conception of logic*

Traditionally, the task of logic has been conceived as that of providing criteria of correctness for inferences by making explicit the rules which are conformed to by correct inferences and violated by incorrect ones; or by characterizing in a general way—*e.g.* by means of skeleton-statements—those propositions which state that one proposition follows from another; and by systematizing these rules and propositions as fully and efficiently as possible. So long as the discoveries of formal logicians were not called in question it was possible to rely on a general inter-subjective logical intuition for the fundamental rules—the so called 'laws of thought'—and for the simplest inferential steps into which complex arguments can be resolved.

It was mainly (i) the attempt to formulate the principles underlying mathematical reasoning, which involved the notion of infinitesimals and later of infinite aggregates, and (ii) the more ambitious logicist attempt to deduce mathematics from logic, which led to an expansion of logic by way of principles whose truth could no longer be based simply on an appeal to logical intuition, especially as the expanded logic led by its own principles to contradiction. The apparent failure to derive mathematics from logic led thinkers to depend—as Kant had done—on such intuitions as were backed up by the particular subject-matter of mathematical construction.

For the formalist the problem is not to expand logic as far as is necessary in order to deduce mathematics from it, but to extract from the whole of logic only so much as is needed in order to reason about formalisms. The formalist is only concerned with what Hilbert calls 'considerations in the form of *thought-experiments* on objects, which can be regarded as concretely given'[1] or what Curry calls, as we have mentioned more than once, demonstrations *ad oculos*. Whereas the logicist had to expand traditional logic for his purposes, the formalists did in some respects contract the logic *in which* they permitted themselves to reason. Indeed Curry does not consider his demonstrations *ad oculos* to be part of logic. He regards mathematics as wholly self-sufficient. The term 'thought-experiment', as used by Hilbert, also seems to imply that in mathematics we observe the outcome of what we are doing when we manipulate objects according to certain rules, rather than draw conclusions merely from statements to statements.

What on this view makes an inference reliable is not a principle of logic—a Leibnizian truth of reason which would be true in every possible world—but the possibility of checking whether the premises imply the conclusion or not, by showing whether in producing the state of affairs described by the premises one is or is not *ipso facto* producing the state of affairs described by the conclusion. On this view '$1 + 1 = 2$' is necessarily true because in producing the juxtaposition of $\langle 1 \rangle$ and $\langle 1 \rangle$ one produces $\langle 11 \rangle$. It is not suggested that those inferences which are capable of constructive checks may not also be found reliable on other, perhaps purely logical, grounds. But the formalist, at least according to the original programme, relies only on constructive tests.

From the point of view of a philosophy of mathematics the relations between an inference and a constructive check upon its validity needs some clarification. It is first of all necessary to note that the construction by which inferences can be checked may be either a construction in fact or a construction 'only in principle'. A constructive check which involves the construction of an integer in the neighbourhood of $10^{10^{10^{10}}}$ is only in principle possible. But even if nothing prevents us from a construction except that we have no means of actually performing it, then the inference is not backed by the construction.

[1] *Op. cit.*, vol. I, p. 20.

The epistemological situation reminds us of a similar issue, the distinction between statements verifiable in fact and those only verifiable in principle. In this instance the difficulty was attacked by developing an epistemological view about the relation between a general law and its verification in perception, which fitted cases of actual verification, and was verbally extended to other cases—the phrase 'in principle', which was not clarified, marking at best an epistemological problem rather than its solution.

But even where a construction in *fact* backs up an inference, the relation is not quite clear. Consider again the constructively backed proposition 'whenever a figure ⟨1⟩ and another such figure are juxtaposed, a figure ⟨11⟩ is produced'. This statement is considered self-evidently true. What is the nature of this alleged non-logical self-evidence? Let us assume that somebody objects to the statement and claims that he had juxtaposed ⟨1⟩ and ⟨1⟩ and not produced ⟨11⟩.

The formalist's answer would be that the objector had not done what he was meant to do; in other words, that his juxtaposition of strokes was not performed *correctly*. But the correctness of a performance is not among its perceptual characteristics; it cannot be, since it is a relation between the performance and an adopted rule—a relation which is more fully expressed by the statement that the performance conforms to the adopted rule. Finding out whether a construction is correct, or conforms to an adopted rule, goes beyond observing what has been—what *happens* to have been—constructed and brings in logical principles and logical reasoning, which though backed by the construction is not validated by it.

Assume that we have adopted a rule *r*, governing constructions (which are, thus, correct if they conform to *r* and incorrect if they violate *r*), and that we assert of a certain construction *c* possessing the characteristic *C* that it conforms to *r because* it possesses *C*. To make this assertion is, among other things, to state or to imply that if *any* construction *x* possessed *C* it would necessarily conform to *r*. This statement is one of logical necessity and may be schematically written as: 'Construction *x* possesses *C*' logically implies 'construction *x* conforms to *r*'. Being hypothetical and general, it is certainly not perceptual; and its self-evidence, if it has any, also cannot be perceptual. It is not a case of 'seeing is believing' because general hypothetical connections and, especially, logical implications are not perceived.

It might be objected that '"Construction *x* possesses *C*" logically implies "construction *x* conforms to *r*"', even if not perceptual, is always trivially self-evident; and that the distinction between constructions *simpliciter* and correct constructions, which introduces the statement of the logical implication, is therefore of little importance. But this is far from true. There are indeed trivial logical implications, *e.g.*: 'Construction x possesses C' logically implies 'construction x conforms to r which prescribes that x should possess C'. But there are others which are not trivial, *e.g.*: 'Construction x possesses C' logically implies 'construction x conforms to r which prescribes that x should possess D'—where the question whether the possession of C by a construction logically implies possession of D turns on the validity of a complicated deduction from 'x has C' to 'x has D', employing *certain admissible principles of inference*. (So-called constructive proofs are on the whole more, and not less, complex than non-constructive.)

The situation then is this: *Prima facie* the formalist does not rely on logical principles but merely on perceptual statements such as 'a given construction of perceptual objects with perceptual characteristics C *ipso facto* possesses characteristics D'. To this the qualification has to be added that the construction has to be correct. The proposition, however, that a construction is correct, *i.e.* that it conforms to an adopted rule, is no longer perceptual but involves a logical implication or an inference the validity of which depends on logical principles. These principles must be adopted before we can decide the correctness of a construction.

In deducing statements about constructions from other such statements one employs fewer logical principles than in classical mathematics. But these principles though suggested by constructions—*e.g.* of strokes and stroke-expressions—are not perceptual judgements. Only if we were to assume that the medium in which we make our constructions is of a special kind so that they can be immediately described by general and necessary propositions without raising the question as to whether a particular construction is correct or incorrect, could we dispense with logical principles. The intuitionists are aware of the fact that ordinary perception is not the medium for such constructions and claim therefore that the general principles of reasoning in mathematics are validated not by constructions in ordinary perception, but in a *sui generis* intuition.

The formalist logic is a minimal *logic*—or better the minimum logic needed for metamathematical reasoning. It is *not* a system of statements describing perceptual features of various constructions. This conclusion is independent of the point urged earlier that mathematical concepts, being exact, differ from perceptual characteristics which are inexact or admit of border-line cases.

VI

MATHEMATICS AS THE ACTIVITY OF
INTUITIVE CONSTRUCTIONS: EXPOSITION

IT is one of the fundamental convictions of the intuitionist school, whose doctrine is the subject of this chapter, that mathematics—if properly understood and practised—is a wholly autonomous and self-sufficient activity. Its methods and insights are regarded as being neither capable of nor in need of the guarantees which the logicists and the formalists each profess to provide. According to the intuitionists the impression that mathematics needs the support of an extended logic or of rigorous formalization has arisen only where mathematics has not been properly pursued.

Logicism and formalism have treated the antinomies of classical mathematics as a malady capable of a cure which would leave classical mathematics substantially intact. The intuitionists consider the antinomies as merely a symptom that mathematics has in many of its branches not been true to itself. Logicism and formalism tried so to reconstruct the building or to secure its foundation that the mathematical work could go on in the upper storeys without much disturbance. The intuitionists attempt to build a new mathematics at all levels by what they regard as the truly mathematical methods.

Both formalists and intuitionists and in particular their modern leaders, Hilbert and Brouwer, acknowledge, as we saw, the influence of Kant's philosophy of mathematics and reject the Leibnizian tradition according to which all mathematical propositions are analytic in the sense that their truth can be demonstrated merely by an application of the principles of logic. Both Brouwer and Hilbert regard mathematical theories as synthetic, in a sense of the term which is based on a mutually exclusive and jointly exhaustive classification of propositions into analytic and synthetic.

Yet Brouwer's conception of the synthetic character of mathematics is very different from Hilbert's, and nearer to Kant. According to Kant, it will be remembered, the axioms and theorems of arithmetic and geometry are synthetic *a priori*—*i.e.* they are descriptive of the pure intuition of space and time and of constructions in it. Brouwer accepts without reservation Kant's doctrine of the pure intuition of time—time apart from any perceptual content—and regards this as the substratum of mathematics. Like Kant he regards such intuition as independent of sense-perception, including in sense-perception in particular the perception of such symbols and operations upon them, as are the strokes and stroke-operations of Hilbert which, together with other marks and operations, constitute the subject-matter of formalist metamathematics.

The subject-matter of metamathematics is *perceptual* objects and constructions, of so simple and transparent a structure that we can be certain of the truth of the synthetic empirical judgements which are descriptive of them. The subject-matter of intuitionist mathematics, on the other hand, is intuited non-perceptual objects and constructions which are introspectively self-evident. Brouwer does appeal, not indeed to the inspection of external objects, but to 'close intro-spection'.[1] The distinction between perceptual and intuitive construc-tions is of some philosophical importance since we can with more plausibility claim that the latter can be apprehended as universal and necessary without the application of the notion of correctness and thus without employing logical principles. (This point was discussed at the end of the last chapter.)

In spite of the differences between the inspectible data of meta-mathematics and the introspectible data of intuitionist mathematics, they have much in common. The most important common feature is that a completed infinite totality can neither be inspected nor intro-spected. In other words neither metamathematics nor intuitionist mathematics can admit statements about actual infinities, only about potential ones.

For a better understanding of intuitionism it is worth asking whether it would reduce to formalist metamathematics if one were to ignore the difference of the substrata, real or alleged, between the two activities. As one would expect, both would employ on the whole the same finite methods—methods such as were described earlier, in our exposition of formalism. However, the formalist would not use them beyond the point at which, having established the consistency of a for-mal system, he could start using it. For the intuitionist, on the other hand, since he cannot find, or hope for, refuge in a formal system, the incentive to use finite methods even in spite of increasing complexity

1 See, *e.g.*, 'Historical Background, Principles and Methods of Intuitionism' in *South African Journal of Science*, Oct.–Nov., 1952, p. 142, footnote.

and difficulty is much greater. Finitist intuitionist mathematics has in fact been developed much further than finitist metamathematics.

Contained in the first chapter of Heyting's *Intuitionism—An Introduction*[1] is a disputation in which one disputant called 'Int' addresses another called 'Form' in the following words: '... you also use meaningful reasoning in what Hilbert called metamathematics, but your purpose is to separate these reasonings from purely formal mathematics, and to confine yourselves to the most simple reasonings possible. We, on the contrary, are interested not in the formal side of mathematics, but exactly in that type of reasoning which appears in metamathematics; we try to develop it to its farthest consequences. This preference arises from the conviction that we find there one of the most fundamental faculties of the human mind.'

For a brief exposition of intuitionism, it will be well first to explain its conception of pure mathematics and the programme based upon this conception; and then to give some examples of the intuitionist method at work especially in dealing with the notion of potential infinity. As to the problem of applied mathematics, the intuitionists have shown even less interest in it than either the logicists or the formalists.

1. *The programme*

Brouwer in one of his more recent English papers[2] describes the situation of the philosophy of mathematics as formulated by the old and new formalists and pre-intuitionists, as he calls those thinkers who in some ways anticipated him, in particular Poincaré, Borel and Lebesgue.

As it presented itself to Brouwer, the situation was this: mathematics, as practised by the pre-intuitionists and formalists, consisted of two separate parts—an autonomous mathematics and a mathematics dependent for its trustworthiness on language and logic. For the autonomous mathematics, 'exact existence, absolute reliability, and non-contradictority were universally acknowledged, independently of language and without proof'. It embraced 'the elementary theory of natural numbers, the principle of complete induction, and more or less considerable parts of algebra and theory of numbers'. The non-autonomous mathematics embraced the theory of the continuum of real numbers. For this a proof of non-contradictory existence was lacking and, as was more or less generally agreed, was needed.

The fundamental theses of the intuitionist philosophy of mathematics are clearly formulated by Brouwer. He describes them as 'two

[1] Amsterdam, 1956.   [2] *Op. cit.*

acts' by which intuitionism 'intervened' in the situation created by its predecessors and the formalists. The acts could also be called 'insights' —a term used frequently by Brouwer. It is best to quote here *verbatim* and at length from his paper.[1]

'The *first act of intuitionism* completely separates mathematics from mathematical language, in particular from the phenomena of language which are described by theoretical logic, and recognizes that intuitionist mathematics is an essentially languageless activity of the mind having its origin in the perception of *a move of time, i.e.* of the falling apart of a life moment into two distinct things, one of which gives way to the other, but is retained by memory. If the two-ity thus born is divested of all quality, there remains the *empty form of the common substratum of all two-ities.* It is this common substratum, this empty form, which is the *basic intuition of mathematics.*'

The doctrine of this and similar passages in Brouwer's writings is substantially that of *The Critique of Pure Reason*—the main difference being that according to Brouwer Kant's intuition of space and the (Euclidean) constructions in it are not part of the intuition which underlies mathematics (see chapter I). Mathematics according to Kant and Brouwer presupposes an intuition which is different on the one hand from sense-perception, of which it is the invariant form, and on the other hand from the apprehension of logical connections between concepts or statements. Just as the experience of, say, climbing a mountain is not to be confused with its linguistic description and communication to others, so the experience of mathematical intuitions and constructions must not be confused with *its* linguistic description and communication (although such linguistic formulation may be of great help to the climber or mathematician and to those who wish to follow his example).

In the same sense in which climbing is not dependent on language, the mathematical activity, with its intuitive insights and constructions, is languageless. According to Brouwer the principles of classical logic are linguistic rules in that those who 'linguistically follow' them may but need not 'be guided by experience'. This means that the rules of classical logic are employed in description and communication but not in the activity itself of constructing; as they are not employed, except as inessential aids, in the activity of mountain climbing. Mathematics is essentially independent, in this sense, not only of language but also of logic.

We must thus according to Brouwer distinguish sharply between two different activities: on the one hand the mathematical construc-

[1] *Op. cit.*

tion; and on the other the linguistic activity, *i.e.* all statements of the results of construction and all application of logical principles of reasoning to these statements. In view of the fundamental difference between the two it makes very good sense to ask whether the logico-linguistic representation is always adequate to the construction; in particular whether the representation does not outrun the construction. That language sometimes outruns its subject-matter is a familiar fact. Usually the danger of its doing so had been regarded as very great in the case of philosophical language and very small in mathematical. But according to Brouwer there is much of it in mathematics too. Thus in the case of all mathematicians who employ the law of excluded middle in reasoning about infinite systems of mathematical objects, language is outrunning and misrepresenting the mathematical reality.

It is again convenient here to quote part of Brouwer's own clear formulation, *verbatim*: 'Suppose that an intuitionist mathematical construction has been carefully described by means of words, and then, the introspective character of the mathematical construction being ignored for a moment, its linguistic description is considered by itself and submitted to a linguistic application of a principle of classical logic. Is it then always possible to perform a languageless mathematical construction finding its expression in the logico-linguistic figure in question?

'After a careful examination one answers the question in the *affirmative* (if one allows for the inevitable inadequacy of language as a mode of description) as far as the principles of contradiction and syllogism are concerned; but in the *negative* (except in special cases) with regard to the principle of excluded third, so that the latter principle, as an instrument for discovering new mathematical truths must be rejected.'

We shall presently consider some mathematical constructions, the examination of which led Brouwer and his followers to reject the law of excluded middle and certain other principles of reasoning for infinite sets of objects. The same rejection we have found in the original limitation of concrete metamathematics by the formalists, who however admit the *formal* application of these principles within the formalized theories of classical mathematics. This way of saving classical mathematics is not open to the intuitionists since it is in conflict with their conception of mathematics as languageless construction.

The limitation of mathematics to the finite methods of formalist metamathematics—whether these be applied to objects of ordinary perception or of intuition—would be a crippling blow to the structure of classical mathematics. But, and this is the second insight of intuitionism, there is a mathematics of the potential infinite, which while avoiding the perceptually and intuitively empty notion of actual, pre-existing infinite totalities, constitutes a firm, intuitive foundation of a new analysis and opens a field of development which 'in several places far exceeds the frontiers of classical mathematics. . . .'

This field of a new autonomous mathematics of the potential infinite is opened by '*the second act of intuitionism* which recognizes the possibility of generating new mathematical entities: first in the form of *infinitely proceeding sequences* $p_1, p_2, \ldots$ whose terms are *chosen more or less freely from mathematical entities previously acquired*, in such a way that the freedom of choice existing perhaps for the first element p_1 may be subjected to a lasting restriction at some following p_r, and again and again to sharper lasting restrictions or even abolition at further subsequent p_r's, while all these restricting interventions, as well as the choice of the p_r's themselves, at any stage may be made to depend on future mathematical experiences of the creating subject; secondly in the form of mathematical species, *i.e. properties supposable for mathematical entities previously acquired*, and satisfying the condition that, if they hold for a certain mathematical entity, they also hold for all mathematical entities which have been defined to be equal to it, relations of equality having to be symmetric, reflexive and transitive; mathematical entities previously acquired for which the property holds are called elements of the species.'

As we shall see in more detail, intuitionist mathematics differs greatly from classical, whether as practised 'naively', as supported by a logicist substructure, or as safeguarded by formalization. Its programme is formulated simply enough, even if its execution involves difficult, or at least very unfamiliar, procedures and concepts and even if the nature of intuitionist construction may not be *prima facie* clear to the non-intuitionist. It is to make mathematical constructions in the medium of pure intuition and then to communicate them to others as clearly as possible so that they can repeat them.

Not every mathematical construction is of equal interest and importance. But there is never much doubt as to which constructions are important, since the motives for finding constructions arise, as in non-intuitionist mathematics, from the curiosity of pure mathematicians and the needs of those who employ mathematics for other purposes. The programme of the intuitionist is to practise intuitionist mathematics, *i.e.* to create or construct mathematical objects since only constructed objects have mathematical existence. It is not to show the legitimacy of these constructions by either logic or formalization. For they are legitimate in themselves, they are self-validating.

2. *Intuitionist mathematics*

To the intuitionist mathematics is the construction of entities in pure intuition, not the promise of such a construction or the enquiry whether it is logically possible.

The classical mathematician, the logicist and formalist allow as legitimate statements to the effect that 'there exists' a number with certain properties although so far no method for constructing this number is known. Such statements—pure existence-theorems—the intuitionist does not allow into his mathematics. He is consequently quite unworried if one finds it odd that a mathematical theorem showing the actual constructibility of some number should only become true after it has been (by his methods) proven. There is no oddity in it to him nor should there be to anybody who understands the intuitionist position, for which 'mathematical existence' means the same as 'actual constructibility'. What is to count as actual constructibility is indeed never quite precisely defined in general terms, but—the intuitionist asserts—it is made clear in practice.

In explaining some of the elementary ideas of intuitionist mathematics—which is all that can be attempted here—I shall be following closely the exposition of Heyting's *Intuitionism—An Introduction*. Heyting leads his reader very much further by explaining the intuitionist approach to special topics of advanced mathematics, such as the theories of algebraic fields and the theory of measure and integration.

Intuitionist mathematics starts, then, with the notion of an abstract entity and of the sequence of such entities. It starts in other words with the sequence of natural numbers. There is no need to formulate a deductive system of elementary arithmetic—for such formulation would be adequate only if it formulated what is self-evident without it. It confers neither self-evidence nor security. It only, at best, reflects it linguistically. For the intuitionist Peano's axioms (see Appendix A) merely formulate self-evident results of the process of generating the natural numbers.

The difference between classical mathematics (equally in its 'naive' and in its logicized or formalized form) and the intuitionist shows itself very clearly when it comes to defining real numbers. In classical mathematics the notion of a real number can be defined in terms of a so-called Cauchy sequence of rational numbers. A classical Cauchy sequence is defined as follows: $a_1, a_2, a_3, \ldots$ or, briefly, $\{a_n\}$ or a, where every term is a rational number, is a Cauchy sequence if for every natural number k (and therefore for every fraction, however small, $1/k$) *there exists* a natural number $n = n(k)$ such that, for every natural number $p, |a_{n+p} - a_n| < 1/k$.

Roughly speaking this means that if we consider any fraction $1/k$ there always *exists* a term, say the nth, such that on subtraction of it from any of its successors, the absolute value of the difference is smaller than $1/k$. (The absolute value of a non-negative number is this number itself, the absolute value of a negative number is that number which results from changing its minus sign into a plus sign.) The absolute value of the difference of pairs of rational numbers thus becomes smaller as we choose them from 'later' members of the sequence.

The definition of the notion of an intuitionist Cauchy sequence can be formulated in almost the same words. The only difference consists in replacing the phrase 'there exists' by the phrase '*there can effectively be found*' or 'there can effectively be constructed'. It is worthwhile to attend to the difference of meaning between these two phrases since it leads to the core of intuitionist mathematics.

Heyting brings it out by means of the following example. Consider the following definitions of classical Cauchy sequences. The first sequence $\{a_n\}$ is: $2/1, 2/2, 2/3, \ldots$ or $\{2/n\}$. In this series each component can be effectively constructed, *e.g.* the thousandth member is $2/1000$. Consider now a second sequence $\{b_n\}$ defined as follows: if the nth digit after the decimal point in the decimal expansion of $\pi = 3 \cdot 1415 \ldots$ is the 9 of the first sequence 0123456789 in this expansion, $b_n = 1$; in every other case $b_n = 2/n = a_n$.

Since the sequence $\{b_n\}$ differs from $\{a_n\}$ in at most one term, it is a Cauchy sequence in the classical sense. But since we do not know of any construction which would show whether or not the critical term occurs in $\{b_n\}$—whether a sequence 0123456789 occurs in π—we have no right to assert that $\{b_n\}$ is a Cauchy sequence in the intuitionist sense. An intuitionist Cauchy sequence, which like $\{a_n\}$ must be constructible, is also called a '(real) number generator'. It is clear that the intuitionist cannot allow the idea of all number-generators into his mathematics—even if it could be shown to lead to no inconsistency in a given formal system.

The identification of the existence with the actual constructibility of number-generators must lead to a thorough modification of the classical notion of the equality and difference of two real numbers. Heyting defines two equality-relations between real number generators, namely 'identity' and (the more important relation of) 'coincidence'. Two number generators $\{a_n\}$ and $\{b_n\}$ are identical—in symbols $a \equiv b$—if for every n, $a_n = b_n$. They coincide—in symbols $a = b$—if for every k we can find an integer $n = n(k)$ such that $|a_{n+p} - b_{n+p}| < 1/k$ for every p.

That we cannot find the required $n = n(k)$ for every k, does not entitle us to say that a and b do not coincide: for an intuitionist negation, just as an intuitionist affirmation, must be based on a construction—not on the absence of a construction. Only if $a = b$ is contradictory, *i.e.* '*only if we can effect a construction which deduces a contradiction from the supposition that $a = b$*', are we entitled to assert that a and b do not coincide, *i.e.* $a \neq b$.

It might be thought that proving in turn that $a \neq b$ is contradictory (impossible) is *ipso facto* a proof that $a = b$. As a matter of fact it is a theorem of intuitionist mathematics that the contradictoriness (impossibility) of $a \neq b$ does amount to $a = b$.[1] But—and this is a very important feature of intuitionist mathematics—'a proof of the impossibility of the impossibility of a property is not in every case a proof of the property itself'. In other words if we write '$\neg$' for 'is contradictory' or 'is impossible'—in the sense in which this notion must be backed by constructive proof—and 'p' for any mathematical affirmation (which is not the affirmation of an impossibility!), then $\neg \neg p$ does not as in classical logic in general imply p. The following example, which shows that this principle is not valid in intuitionist logic, has been given by Brouwer and is also found in Heyting's recent book.

'I write the decimal expansion of π and under it the decimal fraction $\rho = 0 \cdot 333 \ldots$, which I break off as soon as a sequence of digits 0123456789 has appeared in π. If the 9 of the first sequence 0123456789 in π is the kth digit after the decimal point, $\rho = 10^k - 1/3 \cdot 10^k$. Now suppose that ρ could not be rational; then $\rho = 10^k - 1/3 \cdot 10^k$ would be impossible and no sequence could appear in π; but then $\rho = \frac{1}{3}$, which is also impossible. The assumption that ρ cannot be rational has led to a contradiction; yet we have no right to assert that ρ is rational, for this would mean that we could calculate integers p and q so that $\rho = \frac{p}{q}$; this evidently requires that we can either indicate a sequence 0123456789 in π or demonstrate that no such sequence can appear.'

If two number-generators do not coincide (*i.e.* if $a \neq b$) a stronger inequality relation may hold between them. This is the relation of apartness. That 'a lies apart from b'—in symbols $a \# b$—means that 'n and k can be found such that $|a_{n+p} - b_{n+p}| \geq 1/k$ for every p'. It is evident that whereas $a \# b$ entails in general that $a \neq b$, the converse is not true. To the classical mathematician a mathematics which

[1] For the proof see Heyting, *op. cit.*, p. 17.

distinguishes between non-coincidence and apartness in this way would very likely seem unnecessarily complicated and prolix. But this prolixity may be due to mere unfamiliarity. Just as, in philosophy, apparently lucid writers are sometimes confused thinkers, so classical mathematicians may for all their apparent lucidity be fundamentally unclear. Indeed no antinomies have so far been discovered in intuitionist mathematics.

The fundamental operations with real number-generators can be explained in a perfectly straightforward manner. But it must be noted that a real number-generator is not a real number. In classical mathematics one might, having defined a certain number-generator, proceed to define a corresponding real number as '*the set of all* number-generators which coincide with the given number-generator'. But the phrase 'the set of all . . .' does not here refer to a constructible entity and has to be given a new intuitionist content. Indeed to the classical notion of a set there correspond two intuitionist notions, that of a spread and that of a species—a spread being defined by a common mode of generating its (constructible) elements, and a species being defined by a characteristic property which can be assigned to mathematical entities, which have been or could have been constructed *before* defining the species.

In defining a spread the first step consists in conceiving the very general notion of an *infinitely proceeding sequence*, *i.e.* a sequence which can be continued *ad infinitum* no matter how the components of the sequence are determined, whether by law, free choice or what you will. Of such sequences the above defined Cauchy sequences or number-generators are special cases. The intuition of them, and the insight which reveals their mathematical usefulness is—as we have seen (section 1)—claimed to be one of the basic 'acts' of intuitionism.

To the intuitionist the continuum of real numbers is not the completed totality of dimensionless points on a line, but rather the 'possibility of a gradual determination of points'—points describable in terms of the notions of infinitely proceeding sequence and of spread. A spread M is defined by two laws which Heyting[1] whose definitions I am almost literally repeating calls '*spread-law* Λ_M, and '*complementary law* Γ_M'.

A *spread law* is a rule Λ which divides the finite sequences of natural numbers into admissible and inadmissible sequences according to the following four prescriptions, namely

(i) By the rule Λ it can be decided for every natural number k, whether it is a one-member admissible sequence or not.

[1] *Op. cit.*, pp. 34 ff.

(A one-member sequence consists of one natural number, and an n-member sequence of n such numbers. The sequence a_1, a_2, a_3 is called an immediate descendant of the sequence a_1, a_2 and a_1, a_2 an immediate ascendant of a_1, a_2, a_3. And the same terminology is used in the general case of $a_1, a_2, \ldots, a_n, a_{n+1}$ and $a_1, a_2, \ldots, a_n$.)

(ii) Every admissible sequence $a_1, a_2, \ldots, a_n, a_{n+1}$ is an immediate descendant of an admissible sequence $a_1, a_2, \ldots, a_n$.

(iii) If an admissible sequence $a_1, \ldots, a_n$ is given, the rule Λ allows us to decide for every natural number k whether $a_1, \ldots, a_n, k$ is an admissible sequence or not.

(iv) To any admissible sequence $a_1, \ldots, a_n$ at least one natural number k can be found such that $a_1, \ldots, a_n, k$ is an admissible sequence.

The *complementary law* Γ_M of a spread M assigns a definite mathematical entity to any finite sequence which is admissible according to the spread law of M.

Consider now an infinitely proceeding sequence, and subject it to the restriction that, for every n, $a_1, a_2, \ldots, a_n$ must be an admissible sequence in accordance with a spread law Λ_M. Such an infinitely proceeding sequence—briefly *ips*—is no longer a free *ips*; but an admissible *ips* (admissible by Λ_M). The complementary law assigns to each admissible sequence $a_1; a_1, a_2; a_1, a_2, a_3; \ldots$ a mathematical entity—it assigns, say, b_1 to a_1; b_2 to $a_1, a_2; \ldots; b_n$ to $a_1, a_2, \ldots, a_n$. Each of these infinitely proceeding sequences of assigned entities such as $b_1, b_2, b_3, \ldots, b_n$ is called *an element of the spread M*—with b_n as its nth component. Two elements of spreads are equal if their nth components are equal; and two spreads are equal if to every element of one of them, an equal element of the other can be found.

If we understand the notion of spread we can understand the intuitionist notion of the continuum as a possibility of certain actual constructions. Let us—closely following Heyting's exposition as before—consider an enumeration of rational numbers: $r_1, r_2, \ldots$ (*i.e.* we assign to every natural number 1, 2, 3, $\ldots$ —after its construction— a rational number, in a manner which guarantees that no rational number is left out).

We now define the spread M, which represents the intuitionist continuum, as follows: its *spread-law* Λ_M determines that every natural number shall form an admissible one-member sequence; and if $a_1, \ldots, a_n$ is an admissible sequence, then $a_1, a_2, \ldots, a_n, a_{n+1}$ is an admissible sequence if and only if $|r_{a_n} - r_{a_{n+1}}| < \dfrac{1}{2^n}$ ($r_{a_n}, r_{a_{n+1}}$ are the rational numbers which, in our enumeration of rational numbers, have the indices a_n and a_{n+1} respectively). The *complementary law* Γ_M assigns to every admissible sequence the rational number r_{a_n}.

Γ_M thus generates infinitely proceeding sequences of rational numbers. Every such *ips* is an element of M and a real number-generator. Indeed, to any real number-generator c, an element m of M can be found, such that $c = m$. It is worth emphasizing again that nowhere in all this chain of definitions have we assumed an actually given infinity or relinquished the principle that only constructible entities exist.

Just as the notion of a spread does not allow us to assume a completed infinite totality of mathematical entities—being, as it were, a set always in the making but never made—so the notion of a species (a mathematical property) does not allow us to assume actually infinite sets. Obviously the exclusion of 'infinite totality' from mathematics implies the prohibition of properties of infinite totalities.

A *species* is a property which mathematical entities can be supposed to possess. After a species S has been defined, any mathematical entity which has been or might have been defined *before* S was defined, and satisfies the condition S, is a member of the species S.[1] For example, the property of coinciding with a real number-generator is the species 'real number'.

It is important to emphasize with Heyting that the vicious-circle-antinomy (of the set of all sets which do not contain themselves as elements) cannot arise in intuitionist mathematics. For the intuitionist so defines 'species' that only entities which are definable independently of the definition of any given species can be members of that species.

The identification of intuitionist existence with actual constructibility also accounts for fundamental differences between the classical theory of sets or classes on the one hand and the intuitionist theory of species on the other. Thus whereas '$a \in S$' means that a is an element of S—if a is definable independently of S—'$a \notin S$' means that it is impossible for a to be a member of S, in other words that the assumption $a \in S$ leads to a contradiction. Again if T is a subspecies of S (every member of T being a member of S) $S - T$ is not the species of those members of S which are not members of T but of those members of S *which cannot possibly* be members of T. In classical set theory '$T \cup (S - T)$' means the class of all entities which are members of T or of $S - T$ or both and this class is equal to S. In view of the stronger, constructive, definition of $S - T$, the species $T \cup (S - T)$ may but need not be equal to S. (In the former case T is called a detachable species of S.)

[1] Heyting, *op. cit.*, p. 37.

It is clear that the intuitionist theory of cardinal numbers will differ greatly from the classical theory. Thus the requirement of constructibility and the intuitionist conception of negation, as requiring together to be backed up by the actual construction of a contradiction, leads to the denial that a species which is not finite is therefore infinite. (An 'infinite species' is one which has denumerably infinite subspecies, 'denumerable' meaning constructible one-one correspondence with the species of natural numbers.)

3. *Intuitionist logic*

The intuitionist logic is a *post factum* record of the principles of reasoning which have been employed in mathematical constructions. Whereas the logicist formulates these principles in order to abide by them, the intuitionist admits that future mathematical constructions— a notion which to him is unproblematic—might embody principles so far unformulated and unforeseen. Whereas the logicist justifies his mathematics by an appeal to logic, the intuitionist justifies his logic by an appeal to mathematical constructions.

The intuitionist is not concerned with logic in general but only with the logic of mathematics, *i.e.* with 'mathematical logic' in the sense not of a mathematized general logic, but of a formulation of the principles employed in the activity of mathematical construction. Although intuitionists have produced formal systems, which can be made and have been made objects of metamathematical investigation, these systems are regarded by them as linguistic by-products of the 'essentially languageless' activity of mathematics; and as being mainly of pedagogical value.

From a purely formal point of view—that is to say apart from any intended interpretation of the symbols, formulae and transformation rules—intuitionist logic appears as a subsystem of the classical logic. This is particularly obvious in the case of certain formal systems which have been constructed for the purpose *inter alia* of separating intuitionist principles and rules of inference from the wider class of principles and rules which have been adopted by classical and non-intuitionist logicians.[1]

Every intuitionist proposition p, whether or not the (intuitionist) negation occurs in it, is the record of a construction. As Heyting in effect puts it, it says: 'I have effected a construction A in my mind.' An intuitionist negation $\neg p$ is also the record of a construction, and

---

[1] See for example the formal system of Kleene's *Metamathematics*, §§ 19–23, where intuitionistically valid principles, rules of inference and proofs are clearly distinguished from those that are only classically valid.

thus really an affirmation. It says: 'I have effected in my mind a construction B which deduces a contradiction from the supposition that the construction A were brought to an end.' The proposition 'I have not effected a construction...' is of no interest to either the intuitionist or the classical mathematician. But whereas the classical mathematician admits 'there exists a mathematical construction...', even if nobody has so far been able to effect it, such a proposition could from the intuitionist point of view only be an empty promise—perhaps an incitement to research, but not a piece of mathematics.

Considering the intuitionist meaning of p and $\neg p$ we can see at once that if, with the intuitionist, we are to regard mathematics as the science of intuitive constructions then, taking '$\neg$' in its required meaning, the proposition (p or $\neg p$) is not a universally valid principle of the logic of mathematics. By the meaning of the various intuitionist symbols and by the examples of the previous section we see that *if* we adopt the conception and programme of intuitionist mathematics there is nothing at all strange in intuitionist logic. In what follows we shall briefly consider the vocabulary and some theorems of intuitionist logic without attempting a rigid systematization such as would be, in any case, foreign to its spirit.

$p \wedge q$ (p and q) can be asserted if, and only if, both can be asserted; $p \vee q$ (p or q) if, and only if, p or q or both can be asserted. The meaning of '$\neg p$' has been explained already. It is worth noting here that even the strong negation of intuitionist logic has been rejected by some intuitionists as too weak—the reason being that proof of the impossibility of a construction does not seem to them to amount to an actual construction which according to a more radical programme, is alone mathematical. The radical intuitionist requires a completely negationless mathematics and logic. He seems to agree with Goethe's Faust that 'a perfect contradiction remains as mysterious to wise men as it does to fools'.[1]

The intuitionist implication $p \rightarrow q$ is not a truth-function. Heyting interprets it thus: $p \rightarrow q$ can be asserted if, and only if, we possess a construction W which joined to any construction proving p (supposing that the latter be effected) would automatically effect a construction proving q. Or, as he puts it more concisely, a proof of p, together with W, would form a proof of q. We may now put down some intuitionist theorems and non-theorems placing the usual assertion sign $\vdash$ in front of the former and $*$ in front of the latter. Reflection and the meaning of the symbols should ultimately justify the distinction.

[1] For details of this view and references see Heyting, *op. cit.*

(i) $\vdash p \rightarrow \neg\neg p$
* $\neg\neg p \rightarrow p$

(ii) $\vdash (p \rightarrow q) \rightarrow (\neg q \rightarrow \neg p)$
* $(\neg q \rightarrow \neg p) \rightarrow (p \rightarrow q)$

(iii) $\vdash \neg p \rightarrow \neg\neg\neg p$
$\vdash \neg\neg\neg p \rightarrow \neg p$

(In other words, the assertion of the impossibility of p is equivalent to the assertion of the impossibility *of the impossibility of the impossibility of p.* Three intuitionist negations can always be contracted into one.)

(iv) * $p \lor \neg p$
$\vdash \neg\neg (p \lor \neg p)$

(v) $\vdash \neg (p \lor q) \rightarrow \neg p \land \neg q$
* $\neg (p \land q) \rightarrow \neg p \lor \neg q$

In Heyting's formal system $q \rightarrow (p \rightarrow q)$ is an axiom and he gives reasons [1] why he considers it to be intuitively clear. We may observe at this point that at least one intuitionist or near-intuitionist logician denies intuitive clarity to this proposition. Such disagreement about the nature of mathematical intuition is philosophically important and will occupy us in the next chapter.

In developing the usual theory of quantification it is, we have seen, a useful heuristic consideration to regard the universal quantifier as *a kind of* conjunction—and the existential quantifier as *a kind of* alternation-sign. If the members of the conjunction or alternation are finite in number the quantifiers are merely abbreviative devices for the formulation of truth-functional propositions. If the transition to infinite conjunctions and alternations is made, the analogy between universally or existentially quantified propositions on the one hand and conjunctions or alternations on the other, though helpful in some cases, may be very misleading. An 'infinite conjunction' or an 'infinite alternation' are even in the usual theory quite different from a finite conjunction or finite alternation. (See p. 48.)

In developing the intuitionist theory of quantification the heuristic derivation of the principles of quantification from the propositional calculus must be used with even greater care. It must be constantly checked against the principle that mathematical existence is from the intuitionist point of view actual constructibility; and against the particular notions of infinitely proceeding sequences and of spreads, which two notions embody the intuitionist conception of potential

[1] *Op. cit.* p. 102.

infinity. We may again set down the meaning of some of the intuitionist key terms and some theorems and non-theorems.

If $P(x)$ is a predicate of one variable ranging over a certain mathematical species α then

'$(x)\,P(x)$' means that we possess a general method of construction which, if any element a of α is chosen, yields the construction $P(a)$, and

'$(\exists x)\,P(x)$' means that for some particular element a of α $P(a)$ has actually been constructed. By these definitions the following formulae show themselves as theorems or non-theorems respectively.

(vi) $\vdash (x)\,P(x) \rightarrow \neg\, (\exists x)\,\neg\, P(x)$
          $* \neg\, (\exists x)\,\neg\, P(x) \rightarrow (x)\,P(x)$

(vii) $\vdash (\exists x)\,P(x) \rightarrow \neg\, (x)\,\neg\, P(x)$
          $* \neg\, (x)\,\neg\, P(x) \rightarrow (\exists x)\,P(x)$

(viii) $\vdash (\exists x)\,\neg\, P(x) \rightarrow \neg\, (x)\,P(x)$
          $* \neg\, (x)\,P(x) \rightarrow (\exists x)\,\neg\, P(x)$

(ix) $\vdash (x)\,\neg\neg\, P(x) \rightarrow \neg\, (\exists x)\,\neg\, P(x)$

(x) $\vdash \neg\, (\exists x)\,\neg\, P(x) \rightarrow (x)\,\neg\neg\, P(x)$

These sections on intuitionist logic and intuitionist mathematics are of course schematic and incomplete. They can at best convey some of the spirit of intuitionist mathematics. Those interested in making closer contact with its substance are advised to master Heyting's work and refer to its (extensive) bibliography. As to the relation between formalism and intuitionism from the point of view of logic and mathematics readers will find most of the available results in Kleene's *Metamathematics*.

VII

MATHEMATICS AS THE ACTIVITY OF INTUITIVE CONSTRUCTIONS: CRITICISM

In accordance with the plan of this essay we must now examine the intuitionist philosophy of pure and of applied mathematics, and its distinctive theory of mathematical infinity. To the problem of the nature of applied mathematics modern intuitionists have, however, given even less attention than have either the logicists or the formalists. Indeed their philosophy of applied mathematics is something we have largely to conjecture—the basis of the conjecture being chiefly, (i) certain remarks of Brouwer and Weyl (of Brouwer on the affinity of his philosophy to Kant's, of Weyl on the relation between intuitionist mathematics and the natural sciences) and (ii) the reasonable presumption that the intuitionist philosophy of applied mathematics and its philosophy of pure mathematics are consistent with each other. These theories will be treated in the order indicated.

A concluding section will note some indications of new developments springing mainly from a fruitful clash between the formalist and the intuitionist points of view. This section, though expository in character, is best placed at the end of our discussion of formalism and intuitionism as separate points of view.

1. *Mathematical theorems as reports on intuitive constructions*

We have seen that the formalist metamathematician and the intuitionist mathematician make the same claim, that their statements are not statements of logic. They are about a subject matter which is first produced (constructed) and then described. Consequently they are not 'analytic' but 'synthetic'. The constructions of the formalist are made, or can be made, in the physical world; those of the intuitionist in the mind, a medium which is different from sense-perception and open to introspection only. The formalist's statements are synthetic and empirical, the intuitionist's synthetic and non-empirical, *i.e. a priori*.

For the intuitionist every true mathematical statement is justifiable by a construction which is (i) a self-evident experience and (ii) not external perception. He is thus deeply committed to old philosophical doctrines even if he does not wish to discuss them. The intuitionist theory of mathematical truth as validated by self-evident experiences is a restricted version of Descartes' general theory of truth, a theory whose most plausible and mature form is perhaps that given to it by Franz Brentano.[1] The theory of intuitive—non-perceptual—constructions goes of course back to Kant.

If a self-evident experience (or type of experience) is to validate any statement belonging to a public science, it must be intersubjective. It must be capable of being lived through by everybody—at least under suitable conditions. Private experiences, such as are reported by mystics, cannot validate a scientific theory—not even if they are self-evident. Further, the self-evidence of an experience must be intrinsic to the experience or inseparable from it. The person living through the experience must *eo ipso*, without employing criteria, recognize its self-evidence. This implies—as Brentano saw and as Descartes did not always see—that to require such a thing as a 'criterion' of self-evidence is either redundant or fallacious. If an experience is recognized as self-evident in being lived through, no criterion is needed; and if an allegedly self-evident experience is not recognized as such in being lived through, it is not self-evident. Thus 'clarity and distinctness' is a cognate of 'self-evidence'; it is not the name of a criterion of self-evidence. In any case the intuitionists do regard mathematical constructions as intersubjective experiences and their self-evidence as intrinsic.

Yet, although there is no criterion for the presence of self-evidence, there is a criterion for its absence. If two reports about the same intersubjective experience, both linguistically correct, are incompatible, then the experience cannot be self-evident, whatever else 'self-evidence' may mean. For since a linguistically correct report of a self-evident experience is according to the theory necessarily true, and since two linguistically correct reports which are incompatible cannot both be true, the reported experience cannot be self-evident.

Against the view that it is possible for two correct reports of the *same* experience to be incompatible two objections might be raised: first, that no two persons can ever live through the same experience and secondly, that a linguistically correct report of an experience cannot possibly be false. Both objections are reasonable enough, but neither can be raised from the point of view of a self-evidence theory, either of truth in general or of mathematical truth in particular.

[1] See, *e.g.*, *Wahrheit und Evidenz*, Leipzig, 1930.

If no two people can live through the *same* experience, then the experience, not being intersubjective, cannot ever validate the inter-subjective statements of any science. For example, there can be no intersubjective science of introspective psychology, and no intersubjective science of mathematics as reporting intuitive constructions.

Again, if no linguistically correct report of an experience could be false, introspective psychologists and intuitionist mathematicians could make no mistakes except linguistic ones. Yet both introspective psychologists and intuitionist mathematicians do admit the possibility of mistakes which are not linguistic. There must be cases in which introspective psychologists and intuitionist mathematicians would claim to have recognized and amended proper mistakes—mistakes which were not just misdescriptions. A science in which no mistakes can be made and all disagreements are linguistic may not be inconceivable. It is at any rate highly implausible. We shall have occasion presently to discuss some disagreements between intuitionists. It will be seen that they are not regarded by the contestants themselves as linguistic in nature.

Disagreement in reports about one and the same experience may concern its content or merely its self-evidence. Both sorts of disagreement are equally fatal to the claim made for the experience itself, that it is self-evident. To illustrate the first kind, two introspective psychologists or 'phenomenologists' may, after living through the experience of perceiving a certain datum, report differently about it, *i.e.* attribute characteristics to it which are incompatible. The experience then lacks even a clearly demarcated content. As regards the second type of disagreement: of two persons living through the same experience, one may introspect it as self-evident, the other not. I shall be content to appeal only to disagreements of this kind, *i.e.* disagreements about the alleged self-evidence of certain experiences.

The Cartesian philosophical edifice consists in reports of allegedly self-evident experiences, or statements derived from such by allegedly self-evident inferences. There can be no doubt that Descartes' reports on self-evident experiences are incompatible with other reports on the 'same' experiences. One only need live through the experiences to which he appeals in some of his theological and physical arguments and compare one's own reports with his, in order to see that neither his nor one's own are self-evident. The argument from conflicting reports about the same experience plays havoc with Cartesianism.

Can the argument be turned against those self-evident experiences which are supposed to validate the synthetic *a priori* statements of intuitionism? That it can, becomes clear when we reflect on the intuitionist treatment of negation. Heyting, as we had occasion to note already, has described the situation with his usual lucidity. There is a difficulty which Heyting, however, does not consider—the serious difficulty which thus arises for the intuitionist philosophy of mathematics.

Consider the proposition 'A square circle cannot exist'. It is a proposition which Brouwer and Heyting[1] admit as a theorem. It must, therefore, be a linguistically correct report of a self-evident intersubjective experience. Brouwer describes it as a construction which consists in first supposing that we have constructed a square which is at the same time a circle, and then deriving a contradiction from the supposition. But a supposed construction, and one moreover which is unrealizable, is quite different from an actual construction. And although Brouwer reports the self-evidence of the experience which starts with supposing the unrealizable construction, it is not to be wondered at if others report the experience not to be self-evident. And some intuitionists even deny that an unrealizable supposition has 'any clear sense' for them. That the construction is not self-evident is thus proved by the argument from conflicting reports; and a report which is not the report of a self-evident construction is, by definition, not an intuitionist theorem of mathematics.

The same applies to all reports in which intuitionist negation occurs. For, as we have seen earlier, $\neg p$ records—in Heyting's words—the experience of 'having effected in my mind a construction B which deduces a contradiction from the supposition that the construction A were brought to an end'.[2] Since there are conflicting reports—from different intuitionist mathematicians—on this type of experience, no experience expressible only by means of intuitionist negation can be self-evident, and no report on it can be a theorem, in the intuitionist sense, *i.e.* in the sense in which a theorem is a report on a self-evident construction. What applies to reports of the form '$\neg p$' applies also to reports of form '$\neg \neg p$', since $\neg \neg p$ does not imply p.

The argument from conflicting reports on allegedly self-evident experiences is used by the intuitionists themselves against the Kantian claim that the theorems of Euclidean geometry are synthetic *a priori* propositions, being reports on self-evident constructions in the intuitive medium of space as such—space emptied of all sense-content. Brouwer rejects this. But he accepts Kant's claim according to which the theorems of elementary arithmetic are reports on self-evident constructions in time. What for him rules out the synthetic *a priori*

[1] *Op. cit.*, p. 120.   [2] *Op. cit.*, p. 19.

character of Euclidean geometry is not the logical possibility of constructing non-Euclidean geometries, a possibility of which Kant was himself aware; but the disputable self-evidence of constructions which, allegedly, back the Euclidean geometry and no other. The discovery of non-Euclidean geometries may have been among the causes which led to the denial of this self-evidence. It does not by itself imply it.

The role of the argument from conflicting reports on allegedly self-evident constructions, in undermining the security of intuitionist mathematics, is similar to that played by the antinomies in undermining the security of the 'naive' theory of sets and thus of classical mathematics. As regards the antinomies, the main trouble is not so much that they occur as that one can never be sure when and where they will appear again. (See p. 65.) In a similar manner the main trouble caused by the argument from conflicting reports is not its having been successfully applied, e.g., in the case of intuitionist negations and double negations, together with the constructions expressed by them. It is that one can never be sure when and where it will, as it were, strike next. The analogy gains additional weight from the fact that it is one of the aims and claims of intuitionism to banish insecurity from mathematics.

It might be objected that our argument against the intuitionist conception of mathematics has been 'purely philosophical', that it is only a variation of the well-worn argument against the Descartes-Brentano theory of knowledge, which analyses truth in terms of self-evident experiences. This is indeed so. But an argument does not become any the worse for being given a bad name.

It might be pointed out that a slight change could be made in the intuitionist position rendering it impervious to the argument from conflicting reports, a change which would preserve the intuitionist's mathematics, while sacrificing his philosophy of mathematics. All we need to do, it may be suggested, is to concentrate on the intuitionist formalisms which have so far been constructed, develop them further and prove their consistency. The plan is to concentrate on the reports and their mutual consistency, and to forget that they were to be reports of self-evident constructions. This would amount to regarding the intuitionists as formalists who are interested in formalisms of another type than those of the Hilbertians. It is a plan which could be, and has been, followed. But for an intuitionist it means conversion to formalism. The change he is asked to make is fundamental. It is incompatible with his view that mathematics is a languageless activity of self-evident constructions.

140   THE PHILOSOPHY OF MATHEMATICS

If one were to restrict mathematical constructions to those which can be recorded without the use of the intuitionist negation and double negation, intuitionist mathematics would be greatly impoverished, without even being secured against the possibility of conflicting reports. Such security could not be achieved unless a safe class of constructions could be delimited. But this would require a positive criterion of self-evidence and there can be no such criterion. The self-evident is precisely that which neither stands in need of further evidence, nor is susceptible of it.

The *ex post facto* logic of a mathematics which did not presuppose unrealizable constructions would be a so-called 'positive' logic, a logic without negation—*e.g.* the appropriate subsystem of *Principia Mathematicia*. But to define as admissible only those constructions which conform to a positive logic would not be open to the intuitionist. To him the logic of mathematics is validated by self-evident mathematical constructions. It does not validate them. Mathematics to the intuitionist is not only a 'languageless', but also a 'logicless' activity.

It might seem possible to discern within intuitionism a hard core of constructions which can actually be carried out on perceptual objects. The reports on these would then be theorems of a strictly finitist mathematics.[1] But here we come up against a further difficulty, namely that the self-evident constructions take place in intuition and not in sense-perception.

A non-sensory intuition, like a self-evident experience, is a difficult philosophical notion. Though intuitive constructions are alleged to be self-evident, the existence of non-sensory intuitions is by no means uncontroversial. This can be seen by considering the Kantian doctrine of the synthetic *a priori* character of Euclidean geometry, which is disputed by the intuitionists. (See chapter I, section 4.) We are not so much interested here in Kant's argument— from those *non-discursive synthetic a priori* judgements which we are assumed habitually to make, to their ground in a pure intuition of space—as in the properties which the pure intuition in question must have if it is not to be vacuous.

That self-evident constructions must be possible in it, is the most important feature which intuition is supposed to possess. There are, however, two further features which intuition must have, if it is to

[1] For a discussion of the relation between intuitionism and more or less strict variants of finitism see, *e.g.*, G. Kreisel's 'Wittgenstein's Remarks on the Foundations of Mathematics', especially section 6, *British Journal for the Philosophy of Science*, 1958, vol. IX, no. 34.

fulfil its function as the ground of 'the possibility' of Euclidean geometry, namely sharpness and uniqueness. By sharpness I mean that the objects of geometrical constructions must be instances of exact concepts, *i.e.* of concepts which do not have border-line cases. Perceptual-object concepts are inexact. Even if an object is a clear instance of, *e.g.*, 'visual ellipse', this concept, unlike 'geometrical ellipse', has border-line cases. This difference (between exact mathematical and corresponding inexact empirical concepts) which logicism and formalism ignore, in conflating exact and inexact concepts, has already been discussed in the critical chapters on those views. Now the Kantian pure intuition of space is the intuition of instances of *exact* concepts, *e.g.* 'Euclidean point', 'Euclidean line' and the rest. It, so to speak, makes those geometrical objects available. They are not available in sense-perception. (See chapter VIII, where the topic will be treated systematically.)

Sharpness of spatial intuition—the provision of objects for the exact concepts of Euclidean geometry—is not sufficient for Kant's own purpose, which was to show that Euclidean geometry is the *only* geometry whose axioms and theorems are synthetic and *a priori*. For this purpose the spatial intuition must be restrictive, both in so far as it is a place of storage in which objects are recognized, and in so far as it is a place of manufacture in which objects are constructed. It must be so restrictive that the only objects found or constructed in it are Euclidean. For only then can the Euclidean geometry be singled out from among all possible ones as *the* real geometry.

The modern intuitionists reject the Kantian claim that there is a self-evident, sharp and restrictive spatial intuition which alone would make Euclidean geometry a body of *unique*, synthetic and *a priori* statements. (Kant himself was more concerned with showing the synthetic and the *a priori* character of mathematical axioms and theorems than their uniqueness, which latter he was inclined to take for granted.) But the temporal intuition, which they assume, is also self-evident, sharp and unique in the sense that only the objects which are instances of the exact concepts of intuitionist mathematics are constructible in temporal intuition—the objects of other mathematical systems being mere postulations, whose logical possibility must be distrusted even where it cannot be disproved.

The intuitionist account of the theorems of mathematics as reports on self-evident constructions, whatever these latter may be, thus rests ultimately on a self-evidence view of mathematical truth. In view of the serious inroads which arguments from conflicting reports have made into Kant's theory of a pure intuition of space and time and into the modern theory of intuitive constructions—including in particular 'supposed but unrealizable' constructions—modern intuitionism cannot be regarded as a satisfactory philosophy of pure mathematics.

Intuitionism is, however, quite free from the conflation of perceptual and mathematical concepts which we find in the formalist theory of pure, and in the logicist (rudimentary) theory of applied mathematics. It is also not exposed to the objections raised against the logicist claim that mathematics can be reduced to logic, a claim which can be made good only by first defining logic as containing those concepts, statements and inferential rules which are needed to deduce mathematics, as we know it.

2. *Intuitionism and the logical status of applied mathematics*

According to theories of the Frege-Russell type perception and mathematics are connected ultimately through their definition of a natural number as a class of classes whose elements are objects of any kind and thus in particular also perceptual objects. On the other hand according to Hilbert and his followers there is an immediate connection between mathematics and perception. Mathematics in their view is a certain regulated activity of manipulating very simple perceptual objects; metamathematics the theory of such manipulation. I have argued that for each of these theories a special problem of the nature of applied mathematics arises. It arises with special urgency for intuitionism, because of the intuitionist's sharp separation between intuition and perception.

The intuitionist philosophy of pure mathematics leaves room for either of two broad views of applied: on the one hand, the view that applied mathematics must be absorbed into pure, since the theorems of either science are to be taken as reports on self-evident intuitive constructions; on the other hand, the view that applied mathematics is an 'impure', empirical, falsifiable 'mathematics', whose theorems are not reports on self-evident intuitions or constructions at all. Both views are worth consideration. The former goes back to Kant, and is worked out in considerable detail by him in his *Metaphysical Principles of Natural Science*.[1] The latter is expressed as a brief suggestion—almost as an afterthought—by Hermann Weyl in his *Philosophy of Mathematics and Natural Science*.[2]

An outline of Kant's philosophy of pure and applied mathematics as found in the *Critique of Pure Reason* has been given in the introductory chapter. In his later work on theoretical physics, Kant seems

[1] *Metaphysische Anfangsgründe der Naturwissenschaft*, Ak. ed., vol. 4.
[2] Princeton, 1949, Appendix A.

(in order to accommodate this discipline among the *a priori* sciences) to have extended the scope of intuition beyond the field of what is admitted as intuitive in the first *Critique*.[1] If arithmetic and geometry consist of reports on self-evident constructions in time and space, theoretical physics must consist of reports on equally self-evident insights concerning *motion* in space and time. By adding motion to space-time structure, as something into which also we can have self-evident insights, the transition from pure to applied mathematics is made—and it is a transition which according to Kant remains within the field of *a priori* knowledge. (It might be objected that motion presupposes matter and that 'matter' is an empirical concept. There is no need, however, to enter on questions of exegesis.)

Kant, in any case, distinguishes between 'pure natural science', such as theoretical physics, which is possible 'only by means of mathematics' and 'systematic art or experimental doctrine', such as the chemistry of his day, which 'contained no law which would make it possible to represent the motion of chemical parts and their consequences *a priori* and intuitively in space'.[2] Applied mathematics, or what to him amounts to the same thing, *a priori* natural science, is the application (or, as one might also say, the extension) of pure mathematics—arithmetic and geometry—to matter *qua* capable of motion. This extension leads, he argues, to *a priori* or rational physics whose branches are phoronomy, dynamics, mechanics and phenomenology.[3] It is in the sense of these brief remarks that Kant's conception of applied mathematics as a rational natural science must be understood; and it is in the same sense that we must understand his often quoted statement that 'a theory of nature will contain proper science only to the extent to which mathematics is applicable in it'.[4]

It is important to emphasize that Kant does not regard, say, rational dynamics as merely one of many *thinkable* alternative theories, but as part of that one natural science which is synthetic and *a priori*—i.e. which is true of the world and independent of sense-experience. Sense-experience on this view is in no way the ground of our knowledge of rational dynamics but merely the occasion of acquiring it. Just as a child learns that a certain answer to a certain sum is correct on the occasion of experimenting with the beads of an abacus, so Galileo acquired the knowledge of the law of freely falling bodies on the occasion of his experiments at Pisa.

Such a view of applied mathematics may be plausible at a time when only one system of rational dynamics is in existence. The fact

---

[1] See, *e.g.*, his note 2 on p. 482, *op. cit.* [2] *Op. cit.*, p. 471.
[3] *Op. cit.*, p. 477. [4] *Op. cit.*, p. 470.

of there being only one explains to some extent the view that there could only be one. Indeed the conviction that Newtonian dynamics is the only possible dynamics was widely held by physicists for over a hundred years—to be precise for one hundred and one years—after Kant's death. Since then the mere acknowledgement that the special theory of relativity *may* be true and Newtonian physics false makes it impossible to regard the propositions of the latter as self-evident reports on constructions in which pure mathematics is 'applied' to matter, understood as that which is capable of motion.

We now turn to the account suggested by one of the great mathematicians and theoretical physicists of our time, Hermann Weyl. Although himself the author of a 'semi-intuitionist' system, he preferred Brouwer's fully intuitionist system to his own from the point of view of doing more justice to what pure mathematics is or ought to be. (That a theory of pure mathematics *ought* to be intuitionist, follows from Weyl's general philosophical position which is very similar to Brouwer's.)

Intuitionist mathematics in Weyl's opinion is too narrow to accommodate theoretical physics. He is 'disturbed by the high degree of arbitrariness involved . . . even in Hilbert's system'. The alternative seems to him to be found in the work of creative theoretical physicists. 'How much more convincing and closer to the facts are the heuristic arguments and the subsequent systematic constructions in Einstein's general relativity theory, or the Heisenberg-Schrödinger quantum mechanics. A truly realistic mathematics should be conceived, in line with physics, as a branch of the theoretical construction of the one real world, and should adopt the same sober and cautious attitude towards hypothetic extensions of its foundations, as is exhibited by physics.'[1]

Weyl's idea of applied mathematics, in particular of theoretical physics, can be understood as a modification of Kant's conception. He abandons the claim to uniqueness, a claim which he thinks can still be made for intuitionist mathematics. On Weyl's view Newton's dynamics is not, as Kant thought, a report on, or a *description* of, motion in space and time as an invariant feature of our experience of the world, but rather a rational reconstruction of it.

Such a reconstruction is, however, for Weyl not merely occasioned by sense-experience, in particular physical experiment and observation. It must be *in line* with it. And it is always provisional. It is dependent on experimental physics, which implies the possibility, always present, of the emergence of fresh empirical material. This is

[1] *Op. cit.*, p. 235.

always liable to expose a rational reconstruction which seemed to be in line with experience, as having been out of line with it. Newtonian physics had seemed to be in line with experience; but it turned out to be 'less in line' than relativity and quantum-physics.

What, we must ask, does it mean to say that a rational construction is *in line* with experience? Weyl does not explain, and it would seem wrong to require much detailed explanation when all that is put forward is an *obiter dictum*. Yet the concordance, to use a favourite term of Weyl's, which we require to hold between a rational reconstruction and the experience reconstructed is different from that which we require between an empirical generalization and the experience generalized. An empirical law of nature—*e.g.* a general proposition about freely falling physical bodies—must *logically* imply any particular proposition which states evidence for the general law. A 'rational' or mathematically expressed law of nature cannot logically imply the experimental facts with which it is in concordance. Indeed the Galilean law of free fall is a statement about material particles, not physical bodies, and the corresponding Einsteinian law is about metric fields.

Rational reconstructions are formulated in terms of exact concepts which admit of no border-line cases, whereas empirical generalizations about the behaviour of physical bodies, whether within laboratories or outside them, are formulated in terms of inexact concepts. The concepts 'perceptual time-interval', 'perceptual space-interval' and other concepts characterizing perceptual objects are all inexact; whereas the concepts 'Newtonian space-interval', 'Newtonian time-interval', 'Einsteinian space-time interval' and all the concepts of theoretical physics are exact. An analysis of the concordance between the exact statements of theoretical physics (of statements consisting in the application of exact concepts) and inexact perceptual statements requires therefore a preliminary comparison between the logic of exact and of inexact concepts. (See chapter VIII.)

Weyl, by opposing to a pure mathematics, which is separated from sense-perception, a realistic and falsifiable applied mathematics, which serves to describe sense-perception, shows that he is aware of the gulf between empirical and mathematical concepts. But failing to inquire into the logic of the (inexact) perceptual characteristics and compare it with the logic of (exact) mathematical characteristics, he too leaves the philosophical problem of applied mathematics, to the body of which he has contributed so much, almost where he found it.

It might be objected that theoretical physicists solve the problem of applied mathematics *ambulando*—by constructing their theories and making them more and more successful. But no recognition of this obvious fact, and no mere appreciation of the theoretical physicists' work, will do duty for an understanding of the structure of applied mathematics.

3. The intuitionist conception of mathematical infinity

In discussing Hilbert's conception of infinity we distinguished three philosophical positions: finitism, transfinitism and methodological transfinitism. Intuitionism is a moderate finitism which, while repudiating the notion of actual infinities, allows 'reality and intelligibility' to the notion of sequences potentially infinite, *i.e.* growing and incomplete. We saw that each of these positions could be regarded either as a thesis or as a programme; that by holding it as a thesis one implied the falsehood of positions incompatible with it; and that by holding it as a programme one implied that it could be executed or satisfied without necessarily implying the unsatisfiability of programmes incompatible with it.

Now Brouwer regards intuitionism not merely as a programme but also as a thesis; and this particularly in the case of the intuitionist doctrine of potential infinity. He leaves no doubt on the one hand that infinitely proceeding sequences are for him not merely constructions which he prefers to others, or in which he is especially interested. On the contrary he makes it quite clear that infinitely proceeding sequences are the only infinities given to thinking and perceiving beings, and that they are given to them in pure perception or intuition.

On the other hand he leaves no doubt that for him Hilbert's methodological transfinitism is not only incompatible with intuitionism, but is *false*. More than that, he regards it as based on a vicious circle, the eventual repudiation of which is merely a matter of time. He holds that the falsehood of Hilbert's position is revealed by purely reflective experience, an experience 'which contains no disputable elements'.[1]

Pure reflection shows us, so Brouwer holds, that 'the logical (*inhaltliche*) justification of formalist mathematics by a proof of its consistency, contains a *circulus vitiosus*; because this very justification already presupposes the logical (*inhaltliche*) correctness of the statement that the correctness of a proposition follows from its consistency; *i.e.* it presupposes the logical (*inhaltliche*) correctness of the law of excluded middle' This statement of Brouwer's, if true, strikes

---

[1] See, 'Intuitionistische Betrachtungen über den Formalismus', Sitzungsber. preuss. Akad. Wiss., Berlin, 1927, pp. 48–52. Also partly reprinted in Becker's *Grundlagen der Mathematik*, Freiburg, München, 1954, p. 333.

at the very heart of Hilbert's position, both as a thesis and as a programme. At best formalism, in so far as it differs from intuitionism, remains a building up of the stock of mathematical formulae (*des mathematischen Formelbestandes*).

In order to understand Brouwer's criticism, we must remember that Hilbert distinguishes between logical (*inhaltliche*) concepts, statements and inferences on the one hand, and formal—purely symbolic—'concepts', 'statements' and 'inferences' on the other. The law of excluded middle, as applied to actual infinities, is for him a formal law without a logical (*inhaltlich*) counterpart; just as the concept of a transfinite aggregate is a formal concept only. The same is true of other transfinite principles—*e.g.* the axiom that every infinite set can be well ordered—which are formal, but not also logical.

It is essential to Hilbert's programme that the logic or metamathematics used in proving the consistency of a formalism, should be *weaker* than the formalism whose consistency is proved. The logic is weaker than the formalism, if every logical statement has a formal counterpart while not every formal statement has a logical counterpart. Hilbert claims that his logic does not contain the unrestricted principle of excluded middle although the formalism of mathematics does contain it. Brouwer's professed insight is that this logic of Hilbert's does in fact contain the principle implicitly.

Now Brouwer's assertion is remarkable for at least two reasons. First, it was made before Gödel had shown that the weak logic, or metamathematics, which Hilbert intended to employ in demonstrating the consistency of (substantially) the formalism of classical mathematics, is not equal to its task. The framework of metamathematical methods had to be widened. The proposed widening admittedly did not amount to the adoption of the principle of excluded middle. But to adopt the principle of transfinite induction as part of this weak logic or metamathematics is, from the intuitionist point of view, almost as much of an admission of circularity as the adoption of the law of excluded middle would have been. Indeed the stronger one has to make the logic or metamathematics, in comparison with the formalism whose consistency is to be metamathematically proved, the less worthwhile the Hilbert programme must appear to be.

There is a second remarkable feature of Brouwer's report on his self-evident intuition. It concerns not only the circularity of the attempts by Hilbert and his followers to prove without the logical law of excluded middle the consistency of a formalism containing the corresponding formal principle. But it further asserts the circularity of *any* such attempt.

It must, however, be emphasized that Brouwer's insight has not been independently confirmed even by Gödel's proof of the inadequacy of the original metamathematics. It is arguable that metamathematics could be sufficiently strengthened by introducing principles other than the law of excluded middle. It most assuredly is not confirmed with respect to all *possible* metamathematics. Brouwer's insight may well be true insight. Those who do not possess it can only wait till it comes to them; or till, alternatively, further proofs arrive which will show that, contrary to expectation, the difference in strength between given metamathematical systems on the one hand, and the formalisms whose consistency they are intended to demonstrate on the other, is either merely apparent or negligible.

Not everyone is ready to concede the existence of a faculty of intuition, which though different from sense-perception, apprehends particular objects as given. There are many philosophers who hold that introspection does not reveal the presence of any intuitive faculty of the Kantian or Brouwerian type. To deny such a faculty is implicitly to deny Brouwer's positive view that potentially infinite—infinitely proceeding sequences—exist, in the sense of being intuitively constructible. The point here is not to decide whether or not, or in what sense, infinitely proceeding sequences exist, but to show that a statement to this effect is *not* a report on a self-evident, intersubjective experience. Conflicting reports, on the same intersubjective experience, are always enough to show that it is not self-evident. And it is the alleged intuitive self-evidence of such sequences which underlies the claim made for the intuitionist conception of infinity, that it is the only 'real' or 'intelligible' one, and not rather one of a number of alternatives, mathematically equal, though perhaps suitable in varying degrees, for different purposes.

A strict finitist would deny the (constructible) existence of infinitely proceeding sequences, in very much the same manner as the intuitionists deny the existence of actual infinities. *Infinitely* proceeding sequences, he might argue, unlike the *finitely* proceeding, outrun the human capacity for apprehension of the particular. We can imagine a process of stroke added to stroke, up to a point; but there comes a point after which perception and intuition can no longer keep pace. To imagine 'in principle' that the process goes on without stopping, is no longer to imagine. Indeed one can just as little imagine *in principle* as one can perceive (or intuit) in principle. In each case the term marks the transition from apprehending particulars to the entertaining of logically possible but perceptually (and intuitively) empty, general statements.

That intuitionism goes beyond recording what is found or constructible in either intuition or perception (beyond what is *nach anschaulicher Beschaffenheit bestimmt*) was argued by Hilbert and Bernays in the first—pre-Gödelian—volume of their treatise.[1] Moreover, if objection be taken to the intuitionist use of negation and double negation, as implying that unrealizable constructions are as intuitively clear as realizable ones, then precisely the same objection may reasonably be taken against infinitely proceeding sequences.

The intuitionist philosophy of mathematics must be sharply distinguished from the intuitionist mathematics itself. The arguments adduced here against the intuitionist position are directed solely against its philosophy; and in particular against the claim that intuitionist mathematics is not merely one among many possible alternatives but the only one backed throughout by self-evident constructions. In a similar manner the arguments for or against Kant's philosophy of geometry—his singling out of Euclidean geometry as the only one backed by intuition and intuitive construction—do not touch the question of the mathematical merits or demerits of Euclidean geometry.

It is indeed likely that intuitionist mathematics on the lines of Brouwer's programme will continue to flourish, whether his theses are accepted as self-evident insights or not. Many mathematicians are profoundly interested in its problems without being noticeably interested in its privileged status. Belief in the satisfiability of the intuitionist programme has not been shaken. It is not any longer possible to deduce 'mathematics' from 'logic' after the fashion of Frege; or to prove by Hilbert's finite methods that classical mathematics is consistent. It is still possible to pursue intuitionist mathematics as originally conceived.

4. *Interrelations between formalism and intuitionism*

Critics of intuitionism object to a certain vagueness in the intuitionist delimitation of the subject-matter and methods of mathematics. They also object to the intimate connection of intuitionist mathematics with intuitionist philosophy. Yet the mathematics is quite separate from the philosophy—intuitionist mathematical proofs having just the same 'rigour' as the non-intuitionist ones found in the works of classical mathematicians. Moreover, codifications of intuitionist mathematics can be shown, under certain interpretations, to be isomorphic with formalist systems.[2] The main function of the intuitionist philosophy is,

---

[1] *Op. cit.*, vol. I, p. 43.
[2] This question has been investigated by Gödel, Kleene and others. See Kleene, *op. cit.*

150 THE PHILOSOPHY OF MATHEMATICS

as we have seen, to establish a privileged position for intuitionist mathematics. It is taken as being the only 'real', 'proper', or 'intelligible' system of mathematics among an ever growing number of competitors.

Yet the intuitionist philosophy, particularly in its insistence that mathematical existence is constructibility, together with its rejection of the law of excluded middle and of actual infinities, has been of great influence in the development both of mathematics and of the philosophy of mathematics. One frequently finds a desire to combine intuitionist intentions with formalist precision. As a consequence of this mutual interaction, the sharp division of mathematicians and philosophers into logicists, formalists and intuitionists, which was never quite realistic, except for the protagonists of these movements, is likely to lose much of its value and become little more than a pedagogical device.

In reporting on the theory of recursive functions, we saw how it has given precision to the notion of constructive proof in terms of such functions. In this connection we must also mention Kleene's interpretation of intuitionist mathematics in terms of what he calls recursive realizability. He proposes his notion as a precise number-theoretical analysis of the less precise notion of being an intuitionist mathematical theorem.[1]

It is, I think, fair to say—and it is mainly a result of the intuitionists' criticisms of older conceptions, and of their mathematical achievements—that a general scepticism concerning unqualified existence-theorems, unbacked by construction of any sort, is spreading into the farthest corners of mathematics. Some more or less precise justification of existence-theorems is generally being required, or at least found desirable, whenever these theorems concern real numbers or properties of real numbers. The day of the unlimited and unconcerned application of the law of excluded middle, and of the set of all subsets of an infinite set, seems to have passed. To a lesser extent the same holds for the practice of dealing with antinomies by *ad hoc* remedies, such as the theory of types.

A system which, though not as radical as intuitionism, yet shows its influence in many respects, was constructed as early as 1918 by Weyl.[2] Weyl accepts the law of excluded middle for natural numbers (and rational numbers) but not for real numbers or properties of real

[1] Kleene, *op. cit.*, pp. 501 ff.
[2] See *Das Kontinuum*, Göttingen, 1918 and 1932, also later papers, especially 'Über die neue Grundlagenkrise der Mathematik', *Math. Zeitschrift*, 1921, vol. 25; reprinted in Becker's *Grundlagen der Mathematik*.

numbers. The absolute foundation for him of all mathematical constructions is 'the infinite sequence of natural numbers and the concept of existence referring to them'.[1]

The natural numbers in their infinite totality, and statements to the effect that one number is the immediate successor of another or that two symbols represent the same number, form the basis of mathematical construction. Here Weyl agrees with the classical procedure of mathematical analysis. He also agrees with Russell's diagnosis that the definitions of certain concepts in the classical theory of sets, especially the concept of real number, are based on a vicious circle (see p. 46 above). This defect, Weyl insists, must be eradicated, not by *ad hoc* prescriptions or prohibitions, but by the explicit formulation of principles for the actual construction of mathematical entities. The mere definition of a category of objects establishes by no means 'that it makes sense to speak of the objects falling under it as of a determinate and ideally complete totality'. It does not establish that the category is 'denotationally definite' (*umfangsdefinit*).[2]

The positive content of denotational definiteness is determined by Weyl on the one hand by a *provisional* stratification of properties and objects into levels, and on the other by indicating the rules for the construction of second or higher level properties and objects, from those of the first level.

'There is one single fundamental category of objects,' he says, 'the natural numbers; moreover unary, binary, ternary, . . . relations between such numbers. All these we call relations of first level; the category to which such a relation belongs is fully determined by the number of variables (*Unbestimmten*) which it involves.' Next there is the second level. 'The relations of second level are relations whose variables are partly arbitrary natural numbers, partly arbitrary relations of first level. The category to which such a relation of second level belongs is determined by its number of variables, and by the categories of objects, to which each of its variables refers. Relations of third level are those in which there occur variable relations of the second level, etc. To each category K of relations there corresponds a relation $\in (x, x', \ldots; X)$, which means: $x, x', \ldots$ stand in relation X to each other. X is here a variable (*unbestimmte*) relation of category K, and the variables $x, x', \ldots$ refer to the same categories of objects as the variables of the relations X of category K. These $\in$ relations are used together with the relation F of first level (the relation which every integer bears to its successor and only to it) as the initial material (for the constructive process).'[3]

[1] *Kont.*, p. 37.   [2] Becker, *op. cit.*, p. 339.   [3] Becker, *op. cit.*, p. 341.

I shall not repeat Weyl's eight rules for the construction of new mathematical properties from the stock of those available on the first level. Two general points, however, require notice. First, the above described stratification into levels is destroyed by one of his rules. This is the so-called principle of substitution which governs the saturation of propositional functions, such as $R(X, Y)$ where the unsaturated places (*Leerstellen*) refer to sets of level higher than 1 and which permits under certain specified conditions the construction of sets of lower from sets of higher level. Secondly, if $R(x, y)$ is a propositional function, existential quantification—$(\exists x)R(x, y)$—is permitted *only if* the unsaturated place which is to be quantified refers to a natural number or an ordered sequence of natural numbers.[1]

In admitting only sets constructed in a certain way, one of course restricts also the very general notion of a function which has come into use since Dirichlet and Cantor. To show this I shall first give the usual definition of the general concept of function and quote two comments on it, one by Hausdorff, who accepts it without reservation, and one by Weyl who rejects it.

We may start by defining a relation as a set of ordered couples, triplets, . . . , n-tuples. Let R be a set of ordered couples and (a, b) one of them. Two ordered couples, say (a, b) and (c, d), are equal if and only if their first and their second members are respectively equal, *i.e.* if $a = c$ and $b = d$. This implies in particular that $(a, b) \neq (b, a)$ except in the special case in which $a = b$. The set of all first members of the ordered couples of R is called the 'domain', the set of all second members the 'range' of R. (It helps the understanding to think of the first members as x-coordinates and of the second members as y-coordinates.)

In general every first member (every x-coordinate) may correspond to one or more second members (y-coordinates); and every second member to one or more first members. If, however, the correspondence is such that to every second member there corresponds one or more first members, but to every first member there corresponds *only one* second member, then the correspondence is a function 'from' (or 'on') the set of first members 'to' the set of second members. If, in addition, to every second member there corresponds only one first member, then the function is bi-unique or a one-one correspondence. A graph is a good visual analogue of a function—whether or not it is bi-unique. The definition of sets of ordered triplets etc. and of corresponding functions proceeds on the same lines.

In commenting on this notion of a function Hausdorff emphasizes

[1] *Kont.*, p. 29.

that it matters in no way by what rule the correspondence between first and second members is established. It is 'inessential', he says, 'whether this rule is determined by "analytic expressions" or in any other manner, inessential whether our knowledge or the means at our disposal permit the actual determination of $f(a)$ for even a single a'.[1]

The general notion of a function as described above must be abandoned in Weyl's mathematics. Here is his estimate of the situation: 'The modern development of mathematics has led to the insight that the special algebraic constructive principles from which the old analysis proceeded are much too narrow—not only for a logically-natural and general construction of analysis, but also from the point of view of the role which the concept of function has to play in acquiring knowledge of the laws governing material processes. The place of those *algebraic* principles of construction must be taken by general *logical* principles. To renounce such construction wholly, as modern analysis, to judge from the verbal form of its definitions, intends to do (although, fortunately, even here what is said and what is done are two different things), would mean getting utterly lost in a fog. . . .'[2]

Weyl does not claim for his system that it is the only possible foundation for analysis. He claims, however, that his reconstruction is free from vicious circles and 'unnatural' postulates; that its structure is transparent and sufficiently strong to accommodate, in their mathematical formulation, the laws of nature as discovered by contemporary physics. Between the continuum of perception on the one side and the continuum of real numbers as constructed according to Weyl's principles on the other, there still remains a 'deep abyss'.[3] The nature of the continuum of perception—more precisely pure perception or intuition—he considers to be exhibited more nearly by Brouwer's mathematics of which he became one of the most distinguished exponents.[4]

Another important and highly original system which shows the influence of formalism, of its intuitionist critics and of Weyl's work is Lorenzen's operative logic.[5]

The subject-matter of operative mathematics are calculi or formal systems in Curry's sense (see p. 89). As regards the methods of proof and the methods of constructing mathematical objects Lorenzen to some extent goes back to Weyl's early work. His aim is to provide—to use Weyl's words again—a 'logically natural and general construction

[1] *Kont.*, p. 16.
[2] *Kont.*, p. 35.
[3] *Kont.*, p. 71.
[4] See, *e.g.*, Becker, p. 344.
[5] See *Einführung in die operative Logik und Mathematik*, Berlin, 1955.

of analysis'; in his own words, to use 'no unnecessary or arbitrary prohibition' so that the 'methodic framework is left as wide as possible'.[1] He, therefore, does not require that all statements should be effective or intuitionistically true; what he requires is that they should be definite—not denotationally definite, but 'demonstrationally definite' (*beweisdefinit*).

If the figure x is derivable in a calculus K—e.g. in the propositional calculus or in a game of chess played by one person who happens only to be interested in finding out which positions can arise in it—then the statement 'x is derivable in K' is (demonstrationally) definite. So is the statement 'x is not derivable in K' because we know what it is like to derive x in K, i.e. to refute the statement. A rule R is 'admissible' in K if after its addition no more figures are derivable in K than before. Admissibility is thus defined in terms of derivability and underivability, and it is 'definite'. The definition of definiteness is this: '(i) Every statement which is decidable by schematic operations is to be definite. (ii) If for a statement there is determined a definite procedure of demonstration or refutation (*ein definiter Beweis-oder Widerlegungsbegriff*) then the statement itself is to be definite, more precisely demonstrationally or refutationally definite.'[2]

It is not possible here to summarize Lorenzen's work. But it should be noted among other things that he succeeds in replacing Cantor's sets of integers and sets of higher order by a demonstrationally-definite notion—a notion which he uses in a reconstruction of the arithmetic of real numbers and of most of classical analysis. The main means for achieving this is a stratification of language levels reminiscent of Weyl's—the stratification however no longer being provisional. One of the more striking consequences of his procedure is that the difference between denumerable and non-denumerable sets becomes relative only. A set which is denumerable in one level may be non-denumerable in another. (The relative character of the notion of denumerability had been emphasized by Skolem as early as 1922.)

If we forget for a moment the general philosophical positions which in their different ways have inspired the various reconstructions of classical mathematics; and for the moment also forget the question whether all the mathematical systems which have so far been constructed have a common core or any common features to distinguish them as specifically the statements and theories of mathematics, we may be tempted to sum up the situation more or less as follows: Various writers have found themselves dissatisfied with classical mathematics because of its antinomies, because of its lack of rigour,

---

[1] *Op. cit.*, p. 5.   [2] *Op. cit.*, p. 6.

because of this or that other defect. They have put forward various *desiderata* which they felt a mathematical theory should fulfil, and they have set about replacing the old mathematics by a system conforming to these desiderata, duly preserving as much of the old as, consistently with them, could be preserved. Sometimes it has proved necessary to relax the original requirements—as in the case of logicism and formalism; at other times to sacrifice more of the old mathematics renouncing parts of it which had been thought preservable.

We may assume that each mathematician's desiderata (in concept- and statement-formation and in mathematical proof) have either been, as far as possible, satisfied by him or taken by him as satisfiable by some other more competent mathematician. And so, still forgetting our philosophical interest, we might well agree that as an end result many new systems of mathematics have been devised, and many old theories supplied with new foundations. The picture is most reassuring:

> 'There are nine and sixty ways of constructing
> tribal lays,
> And—every—single—one—of—them—is—right!'

Mathematics, we might then conclude, is what all mathematicians do; foundations of mathematics are what some of them work at, and philosophy of mathematics is just reporting these activities with due humility. Such humility, however, has not always been shown by philosophers; and will not be imitated here. In the remaining chapter I shall try to outline a philosophy of pure and applied mathematics by examining the relation of mathematics to perception; and I shall end with some brief remarks on the relations between mathematics and philosophy. To some extent the preceding critical chapters (III, V, VII) have prepared the way for what follows.

VIII

THE NATURE OF PURE AND APPLIED MATHEMATICS

In the preceding chapters we have come across various and mutually incompatible answers to the question 'What is pure mathematics?'. It is logic, says the pure logicist; the manipulation of figures in calculi, says the formalist; constructions in the medium of temporal intuition, says the intuitionist; statements which we abandon more readily than some statements of logic and much less readily than empirical statements, says the logical pragmatist. And there are intermediate positions. The progress of mathematical logic since Boole and Frege has made little difference to the continuation of philosophical disputes about the nature of mathematics.

It may be that the question does not admit of a single and simple answer and misleads us by suggesting such. In a similar way the question 'Why do people obey the law?' suggests that there is one single and simple answer, for example, 'from consent', 'from fear', 'from habit'—each answer being dignified by a high-sounding theory of political obligation. It has been suggested that the answer to the latter question is 'For all sorts of different reasons'; and it might equally be suggested that the answer to the question 'What is pure mathematics?' is 'all sorts of different things'.

A more subtle variant of this same blunt answer has in fact been given by Wittgenstein—for example in the following passage which examines the kind of resemblance which various games, 'language-games' and in particular mathematical 'language-games', bear to each other: 'And the result of this examination is: we see a complicated network of similarities overlapping and criss-crossing: sometimes overall similarities, sometimes similarities of detail.

'I can think of no better expression to characterize these similarities than "family resemblances"; for the various resemblances between members of a family: build, features, colour of eyes, gait, temperament,

156

THE NATURE OF PURE AND APPLIED MATHEMATICS 157

etc., overlap and criss-cross in the same way. And I shall say: "games" form a family.

'And for instance the kinds of number form a family in the same way. Why do we call something a "number"? Well, perhaps because it has a —direct—relationship with several things that have hitherto been called number; and this can be said to give it an indirect relationship to other things we call by the same name. And we extend our concept of number as in spinning a thread we twist fibre upon fibre. And the strength of the thread does not reside in the fact that some one fibre runs through its whole length, but in the overlapping of many fibres.

'But if someone wishes to say "There is something common to all these constructions—namely the disjunction of all their common properties", I should reply: "Now you are only playing with words. One might as well say: Something runs through the whole thread— namely the continuous overlapping of those fibres."'[1]

Here Wittgenstein abandons the search for any characteristic which distinguishes the propositions of pure mathematics from other propositions. One might well agree with him, and still try to find a *common core* in all theories of pure mathematics—some assumption or construction which could be discerned in all of them quite apart from the 'complicated network of similarities overlapping and criss-crossing'. This approach has been made, for example, by Bernays and, as he points out, by Fries before him, whose philosophy is, of course, intimately related to that of Kant. The following passage shows many similarities with passages in Hilbert and Bernays' *Grundlagen der Mathematik*.[2]

'The inquiry into the foundations of mathematics has shown two things. First, that a certain kind of purely perceptual (*rein-anschaulich*) cognition must be taken as the starting-point for mathematics and that indeed one cannot develop even logic as the theory of judgements and inferences without resorting to some extent to such perceptual cognition. What is meant here is the perceptual representation of discrete arrangements (*des Diskreten*), from which we take our most primitive combinatorial representations, in particular that of succession. Constructive arithmetic develops in conformity with this elementary perceptual cognition. We see secondly that constructive

---

[1] *Philosophical Investigations*, translated by G. E. M. Anscombe, Oxford, 1953, §§ 66, 67.
[2] It is taken from 'Die Grundgedanken der Fries'schen Schule in ihrem Verhältniss zum heutigen Stande der Wissenschaft' in *Abhandlungen der Fries'schen Schule, Neue Folge*, Göttingen, 1930, vol. V, 2.

arithmetic is not sufficient for quantitative mathematics (*Grössenlehre*), but that we must for that purpose add certain definite conceptions referring to totalities of mathematical objects (*die sich auf die Totalität von Inbegriffen mathematischer Objekte beziehen*), for example, the totality of numbers and the totality of sets of numbers.'

A mathematical theory consists thus of a hard core of perceptual or, as Bernays is inclined to think, of—in the Kantian sense—intuitive data and constructions, surrounded in some cases by various non-perceptual idealizations referring to ideal totalities.

I have argued above that an 'idealization' of perception is involved in mathematical thinking even before the introduction of infinite totalities. (See especially pp. 60 ff. and pp. 99 ff.) Even such elementary notions as *mathematical* units capable of being *mathematically* added—whether the units and the operation be defined after the fashion of Frege, of Hilbert or of Brouwer—must be distinguished from the corresponding elementary notions of *empirical* units capable of being *empirically* added. The mathematical concepts are exact, *i.e.* do not admit of border-line or neutral cases, whereas the corresponding empirical concepts are inexact. That the exactness of mathematical concepts, statements and theories is an important feature distinguishing mathematical from empirical concepts has, of course, been clearly seen by Plato, and in more recent times—at least as far as geometry is concerned—by Felix Klein among others. That, whatever may be true of empirical concepts, mathematical ones are exact, has been explicitly stated by Frege,[1] and has, so far as I know, been accepted by all philosophers of mathematics, and all mathematicians. To give just one example from a mathematical work chosen at random: 'All that we require for a set *E* to be defined', says the author of a well-known monograph on the Lebesgue integral, 'is that we can say of any given object, whether it is or is not a member of *E*.'[2]

I do not think that philosophers of mathematics have appreciated the relevance of the difference between exact and inexact concepts to the question of the nature of pure and applied mathematics. This is mainly because the logic of inexact concepts has not been given the attention it deserves. Its neglect may be due in turn to the confusing of inexact concepts, which admit of border-line cases, with ambiguous or obscure expressions, whose meaning or use is not determined clearly. Without a clearer view of the logical relations between inexact concepts, the thesis that mathematical concepts are idealizations of (inexact) perceptual concepts must remain much too hazy. To idealize

---

[1] *Grundgesetze*, vol. 2, § 56.
[2] *The Lebesgue Integral*, J. C. Burkill, Cambridge, 1951.

is to idealize something into something else, and unless we know the starting-point as well as the finished product of the operation, the operation itself cannot be clearly understood.

The aim of this chapter is to outline a philosophy of pure and applied mathematics up to the point at which its main theses can be clearly grasped and compared with other philosophical positions. Its more dictinctive features rest on consequences drawn from considering the logical relations between exact concepts, between inexact concepts, and between exact and inexact concepts. I shall, therefore, begin by tentatively exhibiting some simple features of the logic of exact *and inexact* concepts.

As regards pure mathematics I propose to argue that the concepts and statements of any (existing) mathematical theory are—in a precise sense of the terms—purely exact, that they are disconnected from perception, and that in so far as a mathematical theory contains existential statements these, unlike, *e.g.*, empirical and theological statements of existential import, are not unique. As regards applied mathematics I shall, roughly speaking, argue that the 'application' of pure mathematics consists in interchanging perceptual and purely exact statements in the service of a given purpose.

After making the meaning of these theses clearer and arguing in their favour I shall conclude with a brief discussion of the relation between mathematics and philosophy.

1. *Exact and inexact concepts*

For our purpose it is not necessary to enumerate the conditions—if indeed such enumeration be possible—on which a thing can be used as a sign and more particularly as a concept (attribute, predicate, propositional function, etc.). Some general remarks on this question will here suffice.[1]

A thing is used as a sign or, more briefly and less precisely, is a sign, only if correct and incorrect uses of it can be distinguished. This is to say that it must be in principle possible to formulate rules for its use which are capable of being conformed to or violated by the behaviour of a person to whom the intention of conforming to these rules can be imputed. The grounds for the imputation may be such as to justify us in stating either that the person is fully aware of his intention or merely that he behaves as if he had this intention. Between these extremes lie a great variety of intermediate cases. All these possibilities will be covered by saying that the person has adopted the

[1] For a fuller treatment, see *Conceptual Thinking*, Cambridge, 1955, Dover Publications, New York, 1959.

rules and, when there is no need to specify the adopter, that the rules govern the sign.

A sign is a concept only if the rules governing it include a rule of reference, *i.e.* a rule for its assignment or refusal to objects in the wide sense of the term which covers perceptual data, physical things, events, colours, numbers, geometrical patterns—in short everything capable of having signs assigned to it. (The rules of most games, *e.g.* the game of chess, are *not* concept-governing rules.) It is worth emphasizing that this use of the term 'object', and correspondingly of the term 'concept', is—and is meant to be—compatible with all kinds of ontological convictions as to what objects are or are not 'real'. Such ontologies, as we had occasion to observe in comparing Russell's nominalism with Frege's realism as regards number, are usually implemented, though not discovered, by a suitable distinction between complete (categorematic, autosemantic) and incomplete (syncategorematic, synsemantic) symbols; and what follows can be easily reconciled with any ontology and consequent theory of incomplete symbols and their proper contextual definitions. A conceptual system must thus contain concept-governing rules. Whether this is also true of a language in the sense of 'language-game' is not clear to me.

A rule, say *r*, for the assignment or refusal of a sign, say *U*, will here be called an inexact rule of reference, and *U* an inexact concept, if the following two conditions are fulfilled. (i) The first concerns the possible results of assigning or refusing *U* to objects. These are: (a) the case in which assignment of *U* to some object would conform to *r* whereas the refusal would violate it; in which case the object, we shall say, is a positive candidate for *U* and, for the person making the assignment, a positive instance of *U*; (b) the case in which refusal of *U* to some object would conform to *r* whereas the assignment would violate it; in which event the object is a negative candidate for *U* and, for the person making the refusal, a negative instance of *U*; (c) the case in which both the assignment and the refusal of *U* to some object would conform to *r*; in which case the object is a neutral candidate for *U*. For the person who assigns *U* to the object, it is a positive, and for the person who refuses *U* to the object, it is a negative, instance of *U*.

(ii) The second condition concerns the nature of the neutral candidates for the inexact concept *U*. If we define a concept, say *V*, by requiring that the neutral candidates of *U* be the positive candidates of *V*, *V* will again have positive, negative and neutral candidates. (Example: let *U* be the inexact concept 'green' and *V* the concept 'having the neutral candidates for "green" as its positive candidates'. *V* is inexact.) A few remarks on this definition will guard it against misunderstanding. (Nothing in this chapter will turn on the second condition, which, however, seems important for a deeper understanding of the topics here introduced.)

That a concept is inexact, or for that matter exact, is a characteristic of the concept or concept-governing rules; it does not depend on the inventory of the world. The possibility of neutral candidates, and not their actual occurrence, characterizes a concept as inexact. Yet most of the statements which are relevant for our purpose could be replaced by others in which 'inexactness of a concept' is defined in terms not of possible but of actual neutral candidates for it. The controversies surrounding the issue—not always very clear—between 'intensional' and 'extensional' logicians can, therefore, be avoided.

A concept is inexact if both its assignment and its refusal to some object would conform to the rules governing it. That it is not known or that it cannot be known whether a concept is correctly assigned or refused to an object does not make it inexact. In particular, a concept whose assignment or refusal is undecidable by certain permissible methods is not, on that account, inexact. Again, doubt as to whether certain rules for the assignment or refusal of a term should be adopted does not mean that the term is an inexact concept.

For example, one might doubt whether the use of transfinite induction is permitted in securing the consistency of classical number theory; in other words, whether an argument using it is a 'proof' and establishes a 'theorem'. These two terms are used as exact concepts, both by a mathematician who admits and by one who rejects the method of transfinite induction. And the doubt concerns which exact concepts of 'proof' and 'theorem' are to be accepted.[1]

An exact concept cannot have any neutral candidates. For such a concept the distinction between candidates and instances has no point; and clause (c) of the first condition (together with the second condition) has no application. It is possible to define the logical relations between concepts, and the formation of compound concepts by means of connectives, for exact and inexact concepts simultaneously and in such a manner that for exact concepts these definitions reduce to those which are familiar from the logic of exact concepts. In this way one would arrive at a generalized logic of which the logic of exact and the logic of inexact concepts are special cases. Here only the first steps in this direction are taken.

The concepts U and V in whose logical relations we are interested may be exact or inexact. To simplify the discussion we make the reasonable assumption that for every exact concept there is available

[1] See Kleene, *op. cit.*, pp. 476 ff.

a positive and a negative candidate and for any inexact concept in
addition also a neutral candidate. In order to distinguish between the
possible logical relations which may hold between U and V it is convenient to proceed by two steps. The first step is to consider their
relation with respect to their positive or negative candidates only,
i.e. ignoring those objects which are neutral candidates for U or V or
both. This relation will be called their *provisional* relation and written
between square brackets. The second step is to consider the relations
which may hold if also the separate and common neutral candidates
for U and V and their possible elections as positive or negative
instances of these concepts are taken into account. Such relations will
be called *final* and written between curly brackets. (The provisional
relations are, as it were, found before the elections; the final relations
are found after the elections and represent possible results of them.)

The following *provisional* relations can be distinguished:

(i) $[U < V]$, *i.e.* provisional inclusion of U in V, is defined by: U and
V have at least one common positive candidate; and no positive candidate of U is a negative candidate of V. $[U > V]$ is the same as $[V < U]$;
and $[U \leqq V]$ is the same as the conjunction of $[U < V]$ and $[V < U]$.

(ii) $[U | V]$, *i.e.* provisional exclusion between U and V, is defined by:
Each concept has at least one positive candidate which is a negative
candidate of the other and U and V have no common positive
candidates.

(iii) $[U \bigcirc V]$, *i.e.* provisional overlap of U and V, is defined by:
U and V have at least one common positive candidate, and each of the
two concepts has a positive candidate which is a negative candidate
for the other.

(iv) $[U \, ? \, V]$, *i.e.* provisional indeterminacy between U and V, is
defined by: None of the above relations holds.—This possibility cannot arise in the logic of exact concepts.

The possible *final* relations (final inclusion, exclusion, overlap and
indeterminacy) are defined in the same manner as the corresponding
provisional relations—curly brackets taking the place of square
brackets, 'final' taking the place of 'provisional' and 'instance' taking
the place of 'candidate'.

We now inquire into the way in which the provisional relations can
be changed by electing the separate or common neutral candidates for
U and V—if any—as positive or negative instances of these concepts.
Clearly a provisional overlap cannot be changed into a different final
relation. If we have $[U \bigcirc V]$ we must have $\{U \bigcirc V\}$. Again a provisional inclusion cannot be changed into a final exclusion and *vice*

versa. I.e. if [$U < V$], then, no matter what neutral candidates are available for election as positive or negative instances of U or V, we cannot ever have $\{U|V\}$; and if [$U|V$] we cannot ever have $\{U < V\}$. On the other hand a provisional inclusion is compatible with a final overlap. *E.g.* if [$U < V$] (but not also [$V < U$]), if x_0 is a common neutral candidate for U and V, and if x_0 is elected as a positive instance of U and a negative instance of V, then $\{U \bigcirc V\}$. Similarly a provisional exclusion [$U|V$] is compatible with a final overlap $\{U \bigcirc V\}$, *e.g.* if x_0 is a common neutral candidate for U and V and elected as a positive instance of one and also as a positive instance of the other concept. A provisional indeterminacy is compatible with a final inclusion and a final exclusion and sometimes also a final overlap.

The distinction between provisional and final relations can be used in defining the following logical relations between any two—exact or inexact—concepts in the following way:

(i) $U < V$, *i.e.* inclusion of U in V, is defined by: The provisional relation between U and V is inclusion and the only possible final relation is also inclusion. $U > V$ and $U \lessgtr V$ are similarly defined.

(ii) $U|V$, *i.e.* exclusion between U and V, is defined by: The provisional relation between U and V is exclusion and the only possible final relation is also exclusion.

(iii) $U \bigcirc V$, *i.e.* overlap of U and V, is defined by: The provisional relation between U and V is overlap and the only possible final relation is also overlap.

(iv) $U \otimes V$, *i.e.* inclusion-overlap of U and V, is defined by: The provisional relation between U and V is inclusion and two final relations are possible, namely inclusion and overlap. (Similarly for $U \otimes V$.)

(v) $U \phi V$, *i.e.* exclusion-overlap of U and V, is defined by: The provisional relation between U and V is exclusion and two final relations are possible, namely exclusion and overlap.

(vi) $U ? V$, *i.e.* indeterminacy between U and V, is defined by: The provisional relation is indeterminacy. This implies that the possible final relations are inclusion and exclusion and sometimes also overlap. (This possibility, although of little interest for our immediate purposes, is worth consideration.)

If U and V are both exact, then only the relations (i)–(iii) can hold between them and the familiar logical relations between exact concepts find their place in the wider scheme. It is worth emphasizing that (iv) is not an alternation of (i) and (iii) and that (v) is not an alternation of (ii) and (iii).

In defining the above logical relations no restriction has been assumed for the election of neutral candidates and not even the election of the same neutral candidate for a concept U as both a positive and a negative instance of it has been forbidden. Consequently the logical relation of an inexact concept U to itself is $U \otimes U$ and not $U < U$. For if an object is a neutral candidate of U we may elect it once as a positive and once as a negative candidate of U.

That the logical relation between an inexact concept and its complement will not be exclusion is also to be expected for any natural definition of 'complement', which would fit both exact and inexact concepts. Let us say that U and $\bar{U}$ are complements of each other if, and only if, every positive candidate of one is a negative candidate of the other, every negative candidate of one is a positive candidate of the other, and every neutral candidate of one is a neutral candidate of the other. Then the logical relation between an inexact concept U and its complement $\bar{U}$ is $U \phi \bar{U}$ and not $U|\bar{U}$.

The freedom of electing neutral candidates for a concept as its positive or negative instances can be, and is in most conceptual systems, restricted by additional conventions. We must, however, be careful to note two points about them. First, no restrictive convention abolishes neutral candidates of inexact concepts; it merely adds to the rules governing their election and thus to the rules governing inexact concepts. Second, there is a variety of alternative restrictive conventions. Both of these points will be illustrated by the following examples, although only the first is strictly relevant to our present purpose.

Restrictive conventions may be general or special—the general conventions concerning every concept, the special certain concepts only. An example of a general convention is the rule to the effect that if a neutral candidate has been elected as a positive (negative) instance of any concept U, it must not also be elected as a negative (positive) instance of it. An obvious consequence of this convention is that instead of $U \otimes U$, which holds when all elections are independent of previous ones, we have now $U < U$. The general convention, as it were, superimposes inclusion on the original inclusion-overlap.

As an example of a special convention, consider two concepts P and Q, say 'green' and 'blue' which stand in the relation $P \phi Q$. If we add the convention that any common neutral candidate which has been elected as a positive (negative) instance of one of them must be elected as a negative (positive) instance of the other, the result is that the relation $P|Q$ is superimposed on the original $P \phi Q$. The logical necessity of certain modal statements—such as 'Whatever is green is necessarily not blue'—is due to the fact that their denial would violate some adopted special convention restricting the independence on previous elections. Since special conventions are relevant only to the application of inexact concepts, the logic of exact concepts can take no account of such modal statements. This can be done only in a logic of inexact concepts enlarged by suitable restrictive conventions. There is no need here to pursue this topic any further.

Just as the logical relations that are possible between concepts are increased in number by considering the inexact in addition to the exact concepts, so by the same token, the number of possibilities of forming new concepts from already available ones by means of logical connectives is also increased. It is again desirable to define the compound concepts in such a manner that for those that are exact the definitions reduce to the familiar ones. This might be done in the following manner:

The sum of two concepts, say U and V, may be defined by the following stipulation: (a) an object is a positive candidate of $(U+V)$ —in words U or V—if, and only if, it is a positive candidate of either U or V; (b) it is a negative candidate of $(U+V)$ if, and only if, it is a negative candidate of both; and (c) it is a neutral candidate of $(U+V)$ in all other cases. The definition can easily be extended to any finite sum of concepts, and depending on one's attitude, to infinite sums. The sum is defined in terms of candidates for the member-concepts and not in terms of instances and is compatible with general and special restrictions of the independence of the election of neutral candidates into positive or negative instances. If U and V are exact then $(U+V)$ is the familiar sum of exact concepts.

The product $(U.V)$ may be defined as follows: (a) an object is a positive candidate of $(U.V)$ if, and only if, it is a positive candidate of both U and V; (b) it is a negative candidate of $(U.V)$ if, and only if, it is a negative candidate of one or both of them; and (c) it is a neutral candidate of $(U.V)$ in all other cases. The definition can easily be extended to products with more than two members, and it reduces, for exact concepts, to the usual definition. The same is true of the definition of the complement $\bar{U}$ which has been given earlier.

These generalized definitions of sum, product and complement are consistent. Their application yields theorems which are for the most part obvious generalizations of theorems of exact logic. The commutative, associative, and distributive laws are obviously valid. So are the so-called de Morgan laws, e.g. $\overline{U+V}=\bar{U}.\bar{V}$: for by our definitions an object is a positive, a negative or a neutral candidate of $(\overline{U+V})$ if, and only if, it is respectively a positive, negative or neutral candidate of $(\bar{U}.\bar{V})$.

We can, of course, also introduce the 'null concept' 0, of which every object is a negative candidate; and its absolute complement the 'universal concept' $\bar{0}$, of which every object whatever is a positive candidate. If A is any *exact* concept then by our—and the usual—definition of ' $+$ ' ' . ' and '$\overline{}$', $(A . \bar{A})$ represents the null concept and $(A + \bar{A})$ the universal concept. But this is not true in general. For if P is any inexact concept then $(P . \bar{P})$ and $(P + \bar{P})$ have neutral candidates; and these are the same for the sum, the product and for each of their members.

On the whole the proposed definitions conform to customary uses of 'or', 'and' and 'not'. It is possible to define after the usual fashion further connectives, and we can construct a generalized calculus with an interpretation in terms of exact and inexact concepts. For present purposes the preceding definitions and remarks are sufficient.

The compounds of inexact concepts may be exact or inexact. Consider a conceptual system which contains an exact concept A and inexact concepts $P_1, \ldots, P_n$ and is such that the following conditions are fulfilled: (i) that every positive candidate for A is a positive candidate for some P and every negative candidate for A is a negative candidate for every P; (ii) that every neutral candidate for some P is a positive candidate for some other P; (iii) that the positive and negative candidates for the P's respectively exhaust the positive and negative candidates for A. Then $P_1 + P_2 + \ldots + P_n = A$ and the sum of the inexact concepts is itself exact. (Example: 'coloured', which may be assumed to be exact, and its inexact species such as 'green', etc.)

A is an exact concept with inexact species—calling U a species of V if either $U < V$ or $U \otimes V$. By noting in general the exact or inexact character not only of a concept itself, but also of its species, some finer distinctions become possible. The following definitions of *purely exact* and of *internally inexact* concepts will in particular prove useful. A concept is purely exact if, and only if, all its species are exact. (Since every concept is a species of itself, a purely exact concept is exact.) A concept is internally inexact if, and only if, every species of it is either inexact or else has an inexact subspecies. For example, the arithmetical concept 'being a prime number' is purely exact; the concept 'coloured', though in many uses exact, is internally inexact; the concept 'green' is both inexact and internally inexact.

If A is exact, P inexact, and $(A . P)$ not empty, then $(A . P)$ is an inexact species of A (*e.g.*, $A =$ 'being a prime number', $P =$ 'being adored by the Pythagoraeans'). Concepts such as P, and inexact concepts in general, are neither available nor permissible in the systems of Cantor, Frege and their successors including all pure mathematicians.

Indeed these theorists insist, as we have seen, in different words and for different reasons, on that very feature of pure mathematics to which I shall refer by saying that its concepts are purely exact.

Perceptual characteristics which in the philosophical literature are sometimes called 'determinables' or 'respects of likeness', such as 'colour', 'shape', etc., are all internally inexact. In asserting that two perceptual objects resemble each other in a certain respect one is thus applying internally inexact concepts. More particularly, if one perceptual object is to resemble another, say, with respect to the determinable 'coloured', then the objects must be positive or neutral candidates of one or more species of the determinable, e.g. of 'green', 'blue', etc. That statements of resemblance with respect to determinables presuppose the employment of inexact concepts would by itself be sufficient to show that the distinction between exact and inexact concepts is not trivial and philosophically insignificant and that the construction of a generalized logic of exact *and inexact* concepts is worth while.

The intimate connection between the logic of inexact concepts on the one hand and the notions of resemblance between perceptual objects, and of determinable properties on the other, can be brought out in two ways. One may first of all point out that there is a limit beyond which one cannot convey the meaning or use of a perceptual characteristic by defining it in terms of other such concepts. The path from *definiens* to *definiens* will at some point stop, and the need to exemplify one or more *definientia* will arise. To convey the meaning of any perceptual characteristic, say *P*, is among other things to convey, directly or indirectly, a rule to the effect that everything resembling certain objects and not resembling others is to be an instance of *P*. The formulation of such a rule—which I have called 'ostensive rule'— presupposes that the notion of resemblance between empirical objects is clear. It is easily shown that concepts governed by ostensive rules, among others, are inexact.

It is secondly also possible to start at the other end by taking inexact concepts, their difference from exact, and the logical relations between them for granted and to proceed thence to a definition of various notions of determinable and resemblance. Both approaches have their advantages, the first being in any case more obvious and straightforward.[1]

Just as perceptual objects which resemble each other must be

---

[1] I have tried the first approach in *Conceptual Thinking* and the second in *Determinables and Resemblance, Proceedings of the Aristotelian Society*, supp. vol. XXXIII, 1959.

positive or neutral candidates for *inexact* concepts, so mathematical objects which stand in one-one correspondence must be positive candidates for *exact* concepts. Resemblance or empirical similarity is quite different from that one-one correspondence or mathematical similarity, which Frege uses in defining 'number'. In asserting the one-one correspondence between two sets of mathematical objects (say the set of rational numbers and the set of integers) one asserts that every object which is a positive instance of one mathematical concept (say 'rational number') can be matched with some object which is a positive instance of another mathematical concept (say 'integer'). The objects of one set are paired with all objects of the other. The two sets of positive instances are called the 'extensions' or 'ranges' of the two exact concepts (propositional functions, etc.). The extension of an inexact concept, however, is in view of its neutral candidates, which can be elected as positive or negative instances of it, not determined. As was clearly seen by Frege, the 'extensions' of two concepts, one of which is, or both of which are, inexact, cannot be put into one-one correspondence. Frege does not allow inexact concepts. He treats even inexact concepts as if they were exact—as if they had clearly determined extensions.

The distinction between exact and inexact concepts may, as already remarked, seem trivial and philosophically irrelevant. The distinction between resemblance and one-one correspondence is clearly not so. If the second distinction is intimately related to the first, the suspicion of triviality must disappear; and an inquiry into its relevance to an understanding of the nature of pure and applied mathematics can hardly be rejected *a limine*.

The logic of inexact concepts and the generalized logic, of which the logic of exact and inexact concepts are special cases, has here only been begun—by exhibiting the possible logical relations between concepts, by stating some rules for the formation of compounds by means of some connectives, and by defining some new notions in terms of relations between inexact concepts. No more will be needed for the sections to follow. Yet much thought and technical skill will be needed to develop a satisfactory formal system of this logic and changes may have to be made even in the simple beginnings.

2. *Pure mathematics disconnected from perception*

In the last resort all mathematics can be presented in terms of two notions, that of a set or range of an *exact* concept (propositional function, etc.) and that of a function (mapping, etc.) defined in terms of 'set'. This is true as well of classical mathematics as of the later reconstructions of it, which have been discussed in earlier chapters. In the reconstructed systems the notions of set and functions are not abandoned, but only restricted by various qualifications. (See, *e.g.*, Hausdorff's and Weyl's remarks quoted on p. 153.) The concepts of mathematics are thus purely exact, *i.e.* they and all their species are exact. (According to Cantor 'there exist' 2^n subsets of any set of cardinal number n and all of them are exact. In the later systems not all these subsets 'exist'; but those that do are exact and the same is true of the propositional functions or concepts of which these subsets are the ranges or extensions.)

Every perceptual characteristic on the other hand is internally inexact, which, we recall, means that each species of it is either inexact or if exact has an inexact subspecies. An even stronger statement would be defensible, namely, that if P be any perceptual characteristic *all its proper species* (all its species with the exception of P itself) are inexact. But the weaker and less controversial statement will be sufficient for the purpose of comparing and relating mathematics and perception.

That mathematics is purely exact has often been said by logicians and mathematicians in one way or another and will, I think, be generally agreed. It might, however, be objected that the exactness of all mathematical concepts is not of its essence, that it may be a historical accident which will sooner or later reveal itself as such, just as the early concern of mathematics with quantity only revealed itself as an historical accident. The objection is merely verbal. If anything true has been said about mathematics as the science of quantity, it is still true for those theories which fall under the now abandoned definition. Similarly, if anything said about purely exact mathematics is true, it will remain true even if or when a wider notion, including 'inexact mathematics', is generally adopted. In any case whatever inexact mathematics may be, the field of exact mathematics is, and will remain, wide enough to warrant every attention given to it.

The internal inexactness of perceptual characteristics will also hardly be denied by anybody who considers the logical structure of any concept which is exemplifiable in perception. Here it will help to remember that often the same word, *e.g.* ⟨triangle⟩, ⟨addition⟩ and, as we have seen, ⟨natural number⟩ is used for different concepts— namely for (purely exact) mathematical concepts on the one hand and (internally inexact) perceptual characteristics on the other; and to recall the arguments against their conflation in the logicist and formalist philosophy of mathematics.

Yet even if, for the sake of argument, one were to admit perceptual characteristics other than those which are internally inexact, we should still be left with the very wide and interesting field of internally inexact perceptual characteristics. After this concession I shall feel free to omit the qualifications 'internally inexact' when speaking of perceptual characteristics, and 'purely exact' when speaking of mathematical concepts.

One more precaution against misunderstanding: perceptual characteristics are exemplifiable in perception. They are, or fall under, such various categories as 'being a sense-impression', 'being an aspect of a physical object', 'being an aspect of a physical process', etc. Whether or not these categories are legitimate is a metaphysical question; and our use of 'perceptual characteristics' is meant to imply neither a realist, nor a phenomenalist, nor any other metaphysical position.

If mathematical concepts are purely exact and perceptual characteristics internally inexact, the following very simple proposition concerning the relation between exact and inexact concepts and between purely exact and internally inexact concepts acquires philosophical significance:

If $\mathfrak{A}$ is a purely exact concept and $\mathfrak{B}$ is any internally inexact concept, then it follows immediately from the definitions of pure exactness and internal inexactness that neither $\mathfrak{A} < \mathfrak{B}$ nor $\mathfrak{B} < \mathfrak{A}$. Since every species of $\mathfrak{B}$ must by definition be inexact or have an inexact subspecies, $\mathfrak{A}$ cannot be included in $\mathfrak{B}$. Again since every species of $\mathfrak{A}$ must, by definition, be exact, $\mathfrak{B}$ which has inexact species cannot be a species of $\mathfrak{A}$.

Since all mathematical concepts are purely exact and all perceptual characteristics internally inexact, it follows that no mathematical concept includes, or is included in, (is entailed by or entails, is logically implied or logically implies, etc.) any perceptual characteristic. We shall say that mathematical concepts and perceptual characteristics are (deductively) unconnected.

The thesis of the 'unconnectedness' of mathematical and perceptual characteristics, and the characterization of mathematical concepts as unconnected with perceptual, reminds us of Kant's well-known position that mathematical concepts are *a priori*. But Kant assumes his *a priori* concepts, so far as they belong to mathematics, to be instantiated in perception, more particularly, to be characteristics of allegedly invariant perceptual structures, namely space and time. He conceives mathematical concepts as unconnected, not with all perceptual characteristics, but only with those which are *sense*-perceptual. Our thesis is more radical in not recognizing the Kantian distinction between sense-perception and pure perception or intuition. Kant's view further implies a privileged position for certain mathematical theories, namely those allegedly describing pure perception. Lastly Kant does not consider—or does not take seriously—the distinction between exact and inexact concepts.

Plato does distinguish exact mathematical Forms from inexact empirical characteristics. But he too is unaware of the possibility of alternative mathematical systems, and this may be one reason for his metaphysical theory of the Forms. It was of course not open to him to compare the logic of exact and of inexact concepts at a time when the former was in its very first stages.

It is natural and easy to extend the unconnectedness-thesis from concepts to objects, statements and theories. We define a perceptual object as one which has only perceptual characteristics, a mathematical object as one which has only mathematical characteristics; and two objects as 'unconnected' if their characteristics are unconnected. Thus mathematical and perceptual objects are unconnected. We define a statement as perceptual, if and only if to assert it is to assign or refuse a perceptual characteristic to one or more objects; and we define it as purely exact, if and only if the concepts which are assigned or refused in asserting the statement are purely exact. We call the assigned or refused concepts the 'constituent concepts' of the statement; and define two statements as unconnected, if and only if their constituent concepts are unconnected. Mathematical and perceptual statements are thus unconnected. Lastly, we call a theory purely exact, if and only if all the statements and all the constituent concepts of the statements are purely exact; and we call it perceptual if one or more of its statements and, therefore, one or more of their constituent concepts are perceptual. Thus mathematical and perceptual theories are unconnected.

To sum up, we say that pure mathematics and perception are unconnected; or that pure mathematics is disconnected from perception. The latter expression suggests that, in mathematizing perceptual concepts, statements and theories, one so modifies the perceptual concepts that they cease to be perceptual. The modification or idealization amounts, as it were, to a 'disconnection' from perception.

3. *Mathematical existence-propositions*

The problem of mathematical existence-propositions (statements, theorems, metatheorems, etc., but not uninterpreted arrangements of objects belonging to a calculus) is at least as old as the discussion of the so-called mathematicals (see p. 16). Mathematical existence-propositions, *e.g.* 'There exists a Euclidean point' or 'There exists a first natural number', are *prima facie* quite different from other existence-propositions, *e.g.* 'There exists a chair' or 'God exists'. One would expect to find the difference connected with the difference between purely exact and other concepts.

In order to clarify the notion of mathematical existence-propositions it is necessary to clarify to some extent the notion of a proposition. A traditional way would be to characterize propositions (i) as having meaning and (ii) as being true or false. Since 'meaning'—sense, logical content—is possessed also by concepts, the second characteristic serves to distinguish concepts from propositions. To exhibit the meaning of a concept (proposition) is to exhibit the logical relations which it bears to other concepts (propositions). The logical relations between concepts we have to some extent investigated. Those between propositions depend on the rules governing their constituent concepts on the one hand, and, on the other, the rules governing such non-conceptual constituents of the related propositions, as connectives, quantifiers and other operators. The extension of the logic of concepts to include exact as well as inexact ones, will lead also to an extension of the logic of analysed and unanalysed propositions, a task which cannot be undertaken here.

The second traditional characteristic of propositions, their being true or false, is too restrictive. It would exclude rules from propositions, since rules are neither true nor false. The rule, *e.g.*, 'Never to smoke before breakfast'—considered apart from the empirical fact of its being imposed upon, or being adopted by, somebody—is not a concept, it is not true or false and yet it is capable of standing in logical relations to other rules, and is regarded, at least by many people, as a proposition. This is implicitly acknowledged by those logicians who are in the habit of qualifying their use of 'proposition' by adding 'declarative' or 'indicative'.

There is a characteristic, however, which is sufficient to distinguish propositions (including rules) from concepts, namely that the latter, unlike propositions, are capable of being assigned to objects (see p. 160). We, therefore, characterize propositions as (i) capable of standing in logical relations and (ii) not capable of being assigned to objects. Although a proposition may express the assignment of a concept to an object, the proposition—the assignment—is not capable itself of being assigned to anything. This remains true even if one wished, as I do not, to regard propositions as characteristics of 'reality' or of 'the world as a whole'.

Propositions, in our wide sense of the term, can be divided into three classes: (a) logical propositions, which express logical relations between concepts or between propositions; (b) rules, *i.e.* propositions which are capable of being conformed to or violated by the behaviour of their adopters; (c) factual propositions, *i.e.* propositions which are neither rules nor logical propositions.[1] Into this last category fall existential propositions. They are neither logical propositions nor rules. The proposition '*There exists* an object, say x, to which a concept, say P, can be correctly assigned' neither expresses a logical relation between concepts, such as inclusion, exclusion, etc.; nor does it express a logical relation between propositions, such as deducibility or incompatibility. It is a factual proposition.

The characterization of existential propositions as factual is, however, much too loose. 'There exist integers which satisfy Peano's axioms' and 'There exist trees' are very different propositions. Yet they are both factual. In order to characterize the existential propositions of mathematics more closely, we must introduce yet another classification of propositions, namely the dichotomy into, what I shall call, unique and non-unique propositions.

I shall say that a proposition, say p, is unique, if and only if the incompatibility of p with some other proposition, say q, implies that at least one of them is false. Logical propositions are unique. Consider, for example, the incompatible logical propositions $P|Q$ and $P \bigcirc Q$. If one of them exhibits the meaning of P and of Q or, more precisely, if one of them conforms to the rules governing P and Q then the other violates these rules. In this case one of them is true and the other false. The incompatibility of $P|Q$ with $P \bigcirc Q$ admits, of course, also the possibility that neither of them should be true, *e.g.* because $P < Q$. In other words the incompatibility of $P|Q$ with $P \bigcirc Q$ implies that at least one of them is false. As another example consider the logical proposition 'p logically implies q'—which is sometimes and more precisely written as '$p \not{\vdash}_L q$' where the suffix refers to the rules governing the types of conceptual and non-conceptual constituents of a given language or conceptual system. This proposition is incompatible with, *e.g.*, 'p does not logically imply q'. That the two are unique can be seen by an argument which is precisely similar to the preceding one. To assert the internal consistency of a conjunction of concepts or of propositions is to assert logical relations and thus a unique proposition.

Rules, unlike logical propositions, are non-unique. For since rules are neither true nor false the mutual incompatibility of two rules

[1] For greater detail see *Conceptual Thinking*, chapter 3.

cannot imply that at least one of them is false. The rules 'Never to smoke before breakfast' and 'To smoke before breakfast every Monday' are incompatible but their incompatibility does not imply that at least one must be false. The same would hold of the two rules to use ⟨dog⟩ as a label for dogs and to use it as a label for cats, since neither rule is true or false. (What is true or false is that the one is adopted, satisfied, violated, recommended, etc.)

We now turn to factual propositions, *i.e.* propositions which are neither logical propositions nor rules. Particular and general empirical propositions—where 'empirical' may with Popper be understood to mean falsifiable—are clearly unique. Thus of the incompatible particular propositions 'All pieces of copper conduct electricity' and 'There exists a piece of copper which does not conduct electricity' at least one must be false. The last-mentioned proposition is also an example of a unique existential proposition. Again the theological propositions 'Man has an immortal soul' and 'Man has not an immortal soul' are unique—unless we adopt the logical positivists' criterion of meaningfulness.

In order to show that mathematical existence propositions are non-unique, I shall make the following assumption: namely, that a true statement to the effect that there exists an object which has the property P logically implies that P is internally consistent; but that the internal consistency of P does not imply that an object having P exists. Briefly, existence implies consistency; but not consistency, existence. (I also assume the meaning of 'consistency' and its cognates to be uncontroversial, in the sense that any analysis or definition which failed to preserve as true the statement about the relation between existence and consistency, would have to be rejected as inadequate.)

If we compare 'There exists a piece of copper' and 'There exists an immortal soul', on the one hand, with 'There exists a Euclidean point', on the other, we see that the grounds for these existence statements are quite different. Consistency of the constituent concepts is necessary in all cases. But whereas we can make objects for 'Euclidean point' available by decision or postulation, we cannot do this for 'piece of copper' or for 'immortal soul'.

It is legitimate, provided only that 'Euclidean point' is internally consistent, to postulate the existence of Euclidean points, independently of the nature of the physical universe. But it does not follow that in stating 'There exists a Euclidean point' one is stating no more and no less than that 'Euclidean point' is internally consistent. That this would be fallacious follows not only from the general relation between existence and consistency, but from the structure of Euclidean geometry as exhibited, say, by Hilbert or Veblen: if the existence-statements of Euclidean geometry merely expressed the consistency of the concepts of the theory as expressed by its non-existential statements, then it should be possible to eliminate from the theory, once it has been shown to be consistent, all existential postulates without eliminating any of the consequences of the original theory. It would thus be possible to prove the dependence of all existential statements of the theory on its non-existential ones. And this is demonstrably false.

The freedom to postulate the availability of Euclidean points implies the freedom to postulate their non-existence. This means that although the statements 'There exist Euclidean points' and 'There do not exist Euclidean points' are incompatible, this incompatibility does not imply that at least one of them is false. The two propositions, though factual, are, like rules, non-unique. This simple result cannot be expressed in terms of the usual narrow characterizations of propositions, which is bound to obscure it.

The same remarks apply to mathematical existence-propositions in general. Consistency of any purely exact concept—e.g. of 'integer' —permits the postulation of available objects. The various concepts of real number and even of 'integer' as characterized by strict finitists, by intuitionists and by classical mathematicians are quite as different as the concepts of 'point' in Euclidean and in non-Euclidean geometry. We must distinguish as carefully in pure mathematics as in botany or zoology between existence- and consistency-propositions. But whereas the botanist or zoologist cannot create the instances of his self-consistent concepts, the pure mathematician can make the objects of his self-consistent concepts available by his own *fiat*. He not only can do so but continuously does it. Mathematical existence-propositions of the form 'There exists an object such that . . .' are non-unique factual propositions.

The view that mathematical existence-propositions are unique logical propositions is, as we have seen, based on *ad hoc* definitions of what is meant by a 'logical principle'—principles such as the axiom of infinity being classified as logical. Even if our definition of a logical proposition as expressing a logical relation between concepts or between propositions is considered too narrow, no proposition to the effect that objects for an internally consistent concept are available is a logical proposition. Indeed this negative requirement is one of the tests of the adequacy of any definition of 'logical proposition'.

The view that *some* mathematical existence-propositions are unique factual propositions is in the formalist philosophy of mathematics based on the alleged fact that they describe perceptual objects—strokes and stroke-operations. It is due to the confusion of instances of inexact perceptual characteristics with instances of purely exact mathematical concepts. In the intuitionist philosophy the view of the uniqueness of existence-propositions is based on the alleged fact of self-evident, intersubjective, intuitive constructions. This view too has been rejected earlier by means of an old, but efficient, argument.

We have said that the postulation of objects for internally consistent concepts is the foundation of mathematical existence-propositions. This does not imply any answer to the question whether one should actually postulate such objects in a given case. To say it did, would be like taking the fact that the manufacture of a certain type of motor-car is the foundation of statements that motor-cars of this type exist as implying an answer to the question whether one should manufacture them or not. What has been said about existence-propositions in mathematics seems very much in line with the notions of working mathematicians. Indeed their use of the term 'existence-*postulate*' suggests quite clearly the non-uniqueness of mathematical existence-propositions.

It might be objected that not every mathematical theory includes existence-propositions and that all mathematical statements are merely logical statements to the effect that the postulates logically imply the theorems. A mathematical theory would then be merely an exhibition of meaning—of a conceptual network, as it were, without any consideration of whether it 'can catch any objects'. But we may contemplate the conceptual network of (descriptive) zoology with entirely parallel results. We may still raise the question as to how the objects are provided, if any, for the concepts of either system. For zoology the answer will be: by perceptual data or physical objects; for mathematics: by postulation.

It is advisable to avoid saying that every theory of pure mathematics contains existential propositions which are non-unique, and to say rather that the existential propositions which are either contained in the theory or else by which it can be supplemented have this characteristic. With this understanding we can sum up the argument of the last two sections thus: every theory of pure mathematics—which is formulated in terms of sets and functions or cognate concepts—is purely exact and existentially non-unique.

4. *The nature of applied mathematics*

Pure mathematics is (logically) disconnected from perception. Yet in applied mathematics, particularly in theoretical physics, pure mathematics and perception are brought together. What is the nature of this relationship? The correct ground for an answer seems to have been largely prepared by the preceding discussion.

It is convenient to start the discussion by quoting a concise statement by a theoretical physicist. The statement is similar in spirit to others to which we have had occasion to refer, from working mathematicians and scientists employing mathematics in their field; and their number could be increased indefinitely. P. A. M. Dirac[1] is pointing out that quantum mechanics needs a different mathematical apparatus for its formulation from the one used in classical physics, because the physical content of the new ideas 'requires the states of a dynamical system and the dynamical variables to be connected in quite strange ways that are unintelligible from the classical standpoint'. He proceeds to express, as follows, his general view of the structure of quantum mechanics and, it appears, of any physical theory:

'The new scheme becomes a precise physical theory when all the axioms and rules of manipulation governing the mathematical quantities are specified and when in addition certain laws are laid down connecting physical facts with the mathematical formalism, so that from any physical conditions equations between the mathematical quantities may be inferred and vice versa. In an application of the theory one would be given certain physical information which one would proceed to express by equations between the mathematical quantities. One would then deduce new equations with the help of the axioms and rules of manipulation and would conclude by interpreting these new equations as physical conditions. The justification for the whole scheme depends apart from internal consistency on the agreement of the final results with experiment.'

Dirac in speaking of 'laws laid down connecting physical facts with the mathematical formalism' is careful not to prejudge the nature of the connection in terms of one-one correspondence between physical and mathematical characteristics or objects, or in terms of an instantiation of mathematical characteristics by physical objects.

Eddington, who considers the logical structure of the theory of relativity in his equally classical work,[2] is, it seems to me, less clear. Having 'developed a pure geometry, which is intended to be descriptive of the relation-structure of the world', he formulates what he calls the 'principle of identification' as follows: 'The relation-structure presents itself in our experience as a physical world consisting

---

[1] *The Principles of Quantum Mechanics*, 3rd ed., Oxford, 1947, reprinted 1956, p. 15.
[2] *The Mathematical Theory of Relativity*, 2nd ed., Cambridge, 1924, p. 222.

of *space, time* and *things.* The transition from geometrical to the physical description can only be made by identifying the tensors which measure physical quantities with tensors occurring in the pure geometry; and we must proceed by inquiring first what experimental properties the physical tensor possesses, and then seeking a geometrical tensor which possesses these properties *by virtue of mathematical identities.'*

The difficulty lies in the meaning of the term 'identification'. If what is meant is merely that for certain purposes physical characteristics are treated *as if* they were mathematical, then Eddington's remarks would be quite similar in intention to Dirac's. If what is meant is that the identity of mathematical and physical characteristics is discovered, conjectured or postulated, then the principle of identification is false, because incompatible with the disconnectedness of mathematics from perception. Eddington's later writings, in particular his explicitly philosophical ones, tend to confirm the second interpretation—which ignores the fundamental difference between purely exact concepts and definite correspondences as found in pure mathematics on the one hand, and inexact concepts and resemblances as found in perceptual propositions, on the other.

In order to explain the relation between mathematical and perceptual characteristics in theoretical physics and applied mathematics in general it is fortunately not necessary to consider any mathematically complex theory such as quantum-mechanics or the theory of relativity. It can be considered 'without loss of (philosophical) generality', in terms of the extremely simple example which served us in our criticism both of the logicist and the formalist view of applied mathematics. Consider once again, but now in the light of the more precise and detailed discussion of the logic of exact and inexact concepts, the propositions

    (1) '1 + 1 = 2'
    (2) 'One apple and one apple make two apples'.

Proposition (1) is a proposition of pure mathematics which can be analysed in many different ways, *i.e.* it may be regarded as a proposition belonging to various arithmetical theories, which need not be consistent with each other—differing for example in their transfinite postulates. (By saying that two such theories are inconsistent, I mean that the set of postulates of the two theories, taken together as jointly defining not different number-concepts, but the *same* number-concept, is an inconsistent conjunction.) Yet each of these versions involves only purely exact concepts. Thus, the concepts of (1) and

(2) may in Fregean fashion be regarded as exact concepts characterizing units and couples; in the fashion of Hilbert as exact concepts characterizing strokes on paper (which strokes, however, are to be considered not as positive instances of inexact concepts which do, but of exact concepts which do not, admit of neutral candidates); or lastly in the fashion of Brouwer as exact characteristics of self-evident, intuitive constructions. Each of these versions leaves room for further more detailed differences in the rules governing these exact concepts—in accordance with actual or possible variants of these principal doctrines in the philosophy of mathematics. The same is true of the analysis of mathematical addition. Frege regards mathematical addition as the purely exact relation of the logical sum (of exact ranges of exact concepts); Hilbert as an exact concept characteristic of the juxtaposition of ideal strokes; and lastly Brouwer regards it as an exact characteristic of an intuitive counterpart of this perceptual operation.

Proposition (1), however analysed, thus involves only purely exact concepts and is disconnected from perception. It certainly, according to Hilbert and Brouwer, implies that the concepts involved are not empty. It is existential and, as we have argued, non-uniquely so.

Proposition (2) may be analysed in two different ways. It may first of all be regarded as being also purely exact. This, as we have seen, the logicist does by regarding (2) as a substitution instance of (1)—formed by substituting specified unit-classes of apples for unspecified unit-classes in (1) and their logical sum for the unspecified logical sum in (1). The transition is for non-empty, distinct x and y, from

$$(x)(y)(((x \in 1) \& (y \in 1)) \equiv ((x \cup y) \in 2))$$

to

$$((x_0 \in 1) \& (y_0 \in 1)) \equiv ((x_0 \cup y_0) \in 2).$$

In the formalist system, (2) is interpreted as being isomorphic with (1)—apples and their juxtaposition being tacitly considered as instances of exact concepts. I shall here refer to any transcription of (2) which turns it, in the way indicated, into an exact proposition as (2a).

Proposition (2) can, however, also be regarded as an empirical statement of the result of some physical addition of physical objects. The concepts 'physical unit', 'physical addition', 'physical couple' —in the various senses of these terms—are all internally inexact concepts; and the proposition, of which these inexact concepts are constituents, is accordingly internally inexact. 'One apple and one apple make two apples' is an empirical law of nature, which, unlike '$1+1=2$', is capable of being confirmed or refuted by experiment and observation. I shall refer to any analysis of proposition (2) which turns it into a general empirical proposition involving internally inexact (perceptual) characteristics as (2b).[1]

It has been argued, in the critical chapters above, that to consider the relation between (1) and (2a) is not even to touch the problem of what is involved in the application of pure mathematics to experience. To give an account of the relation between (1) and (2b) in terms of that between (1) and (2a) is to make the mistake of confusing purely exact concepts, their objects and the propositions involving them with internally inexact concepts and their objects and the propositions involving them. It is to disregard the fundamental difference between the logic of exact and inexact concepts, and their (logical) disconnectedness.

Since (2a) and (2b) are of an entirely different structure and since therefore (2b) is neither an instantiation of (1), nor isomorphic with it, the 'application' of (1) which results in (2b)—the idealization or mathematization of (2b) by (2a)—consists in *replacing* (2b) by (2a). This replacement is justified by the purpose in hand. In particular, if (2a) serves together with other mathematical propositions as a premiss for the deduction of further mathematical propositions, and if some of these can be considered as idealizations of new empirical propositions, the original replacement of (2b) by (2a) is justified as an aid in discovering new empirical truths. The procedure of theoretical physics and of applied mathematics in general is to *replace* empirical propositions by mathematical ones, to deduce mathematical consequences from the mathematical premisses, and to replace some of these consequences by empirical propositions. That this procedure can be, and has often been, highly successful, depends on the world's being what it is. That satisfiable rules governing—more or less strictly—the interchange of exact and inexact concepts (before and after mathematical deduction) have been found, depends on those features of the world which go under the name of human ingenuity.

The difference between the applied mathematics of apple-addition on the one hand, and quantum-mechanics and relativity-physics on the other, is only a difference of complexity. In quantum-mechanics and relativity-physics, two successive interchanges of purely exact and internally inexact concepts or propositions (physical concepts or propositions first being replaced by mathematical and afterwards mathematical by physical) are usually separated by long chains of mathematical reasoning, whereas in the physics of apple-addition the

[1] For an analysis of empirical laws of nature, see *Conceptual Thinking*, chapter XI.

intervening chain of mathematical reasoning may be small or non-existent. Moreover, whereas in applying the pure geometry of Hilbert space to the physical phenomena of atomic physics, and in applying the calculus of tensors to the physical phenomena of moving bodies, *not* every mathematical concept or proposition is paired with a physical, the pairing of mathematical with physical concepts or propositions in our apple-example is complete.

It might be argued that frequently, before pure mathematics can be applied to sense-experience, it must first be extended by the introduction of new concepts and postulates governing their use. Thus according to Russell[1] pure mathematics is extended into rational dynamics by the introduction of such concepts as 'mass', 'velocity', etc. and corresponding new postulates.

Even if we grant the doubtful possibility of distinguishing sharply between these and purely logical or mathematical concepts, the concepts of rational dynamics are purely exact. 'Mass' and 'velocity', as used in rational dynamics, are deductively unconnected with those concepts of mass and velocity which are characteristic of sense-experience, within and outside laboratories, and which like all empirical concepts are internally inexact. (Rational dynamics comprises no concepts admitting of border-line cases.) In other words the concepts of rational dynamics stand to their empirical counterparts—if any—in the same relation of unconnectedness as does, for example, 'physical addition' in its various senses to 'mathematical addition', whether this concept is defined after the logicist, formalist, intuitionist or any other fashion as a purely exact concept.

In this context we must mention a distinction between concepts of pure and applied mathematics, which is due to Karl Menger.[2] He defines a quantity as an ordered pair whose first member is an object and whose second member is a number. Two quantities are consistent unless they have the same object and different numerical values. If the object is not a number, but, for example, a physical distance or an act of reading a scale, then the quantity belongs to applied rather than to pure mathematics. A class of mutually consistent quantities is briefly called a 'fluent'. If the first members of its elements are numbers the fluent is a function of pure mathematics. If the first members of its elements are not numbers the fluent expresses a relationship in applied mathematics.

Menger's penetrating analyses in terms of these key-concepts have

---

[1] *Principles of Mathematics*, 2nd edition, London, 1937, pp. 465 ff.
[2] See *Calculus—A modern approach*, Boston, 1955, and various papers mentioned there.

come to my notice too late to be given the attention they deserve. I must therefore be content to remark that his fluents, in particular those belonging to applied mathematics, are purely exact and thus deductively unconnected with empirical concepts, which are internally inexact.

To sum up our discussion of applied mathematics: the 'application' to perception of pure mathematics, which is logically disconnected from perception, consists in a more or less strictly regulated activity involving (i) the replacement of empirical concepts and propositions by mathematical, (ii) the deduction of consequences from the mathematical premises so provided and (iii) the replacement of some of the deduced mathematical propositions by empirical. One might add (iv) the experimental confirmation of the last-mentioned propositions—which, however, is the task of the experimental scientists rather than the theoretical.

The view put forward agrees substantially with Dirac's statement, with the views of v. Neumann (see p. 60), and with many others. And, as was pointed out earlier, it also has affinities to Curry's views on applied mathematics. The novel feature of the present account is the exhibition of the contrast between (purely exact) mathematical and (internally inexact) empirical concepts and propositions, a contrast which is at its clearest in the simple theorem about their disconnectedness.

The 'application' of mathematics in theoretical physics is understood by some contemporary philosophers in a rather different way. They hold that by mathematical reasoning empirical conclusions are *directly* deduced from empirical premises, without the interchange of exact and inexact concepts and propositions before and after the mathematical deduction. This is suggested, for example, by the well-known aphorism of Benjamin Peirce that mathematics is 'the science which draws necessary conclusions'. It is implicit also in Kant's philosophy of mathematics. It may be implied too—or is at least not rejected—in some modern writings, *e.g.* R. B. Braithwaite's excellent *Scientific Explanation*.[1] But to ignore the interchange of exact and inexact concepts in the arguments of theoretical physics, is to extend the conflation of mathematical and empirical concepts from the philosophy of mathematics to the philosophy of science.

5. *Mathematics and philosophy*

Mathematicians are engaged in filling two kinds of gap, not always sharply distinguishable—gaps consisting in the absence of theorems

[1] Cambridge, 1953.

within existing theories, and gaps consisting in the absence of theories. Philosophical considerations are more likely to be influential when the task is not so much to find theorems, as to find theories. Again, they will be more influential in the construction of theories intended to provide the 'foundations' of mathematics, than in theories providing the mathematical apparatus, say, for a particular branch of physics. That at least the originators of logicist, formalist and intuitionist mathematics have been strongly influenced by philosophical assumptions, insights, prejudices (whatever one may call them) cannot be doubted if we take their own statements seriously. In order to see the relation between mathematics and philosophy more clearly, one may well consider a little more closely the useful and widely adopted distinction between analytic philosophy and metaphysics. A somewhat schematic and over-simplified treatment will be sufficient.[1]

Analytical philosophy tended to be regarded at one time as the exhibition of the 'meaning' of common-sense statements and of statements and theories belonging to special fields of inquiry—such exhibition being regarded not as changing but merely as bringing into clear view what was meant. After the widespread adoption of Wittgenstein's advice, to look not for the meaning but for the use of linguistic expressions, his followers regarded their analytical philosophy as the exhibition of the rules governing the linguistic expressions of the analysed beliefs and theories. The requirement that analysis must not change what is analysed was still acknowledged. Wittgenstein formulates it by saying that 'philosophy may in no way interfere with the actual use of language; it can in the end only describe it'.[2] I call this type of analysis 'exhibition-analysis'.

Without going into the question to what extent exhibition-analysis is a fruitful philosophical method, it seems obvious that not all philosophy, not even all that is pursued under the name of analytic philosophy, is exhibition-analysis. Analytical philosophers and others often find it necessary to go beyond the exhibition of rules and to change them; in particular to preserve only some of the rules as they are, while replacing others by more suitable ones, the suitability depending on various circumstances and purposes. Thus it might quite plausibly be held that the set theoretical antinomies have been brought to light by an exhibition-analysis of classical mathematics (and perhaps even of common-sense beliefs); and that the

---

[1] For more detailed accounts see *Conceptual Thinking*, especially chapters XXX-XXXIII, and *Broad on Philosophical Method* to appear in the Broad-volume of the *Library of Living Philosophers*, ed. Schilpp.

[2] *Philosophical Investigations*, Oxford, 1953, p. 49.

mathematical and philosophical problems resulting from this discovery
include the problem of a suitable replacement of some rules governing
the term 'set' and its cognates in classical mathematics and common
speech by others. I call this type of analysis 'replacement-analysis'.

Replacement-analysis thus consists in replacing a defective analysandum by a sound analysans—a defective set of rules by a sound one
—provided, of course, that the analysans and the analysandum have
enough in common to justify one in speaking of an analysis at all. If
one is to know when a replacement-analysis has been successful, one
has to agree upon (i) some more or less clear criteria of soundness and
(ii) a relation which must hold between analysans and analysandum.
A problem in replacement-analysis has thus the following general
form: given a criterion of soundness of rules governing concepts and
other propositional constituents and given an analysing relation—to
replace a conjunction of defective rules by a conjunction which is sound
and which stands in the analysing relation to the defective set. The
criteria of soundness and the analysing relation which are presupposed
in making a replacement-analysis may, and do, greatly vary both in
their content and in the degree of precision with which they are formulated; here one man's meat may well be another man's poison. The
question, how the choice between different criteria is to be justified,
arises immediately.

Neither exhibition- nor replacement-analysis can justify the
choice. The exhibition-analysis will, if correct, only show which
choice has been made; the replacement-analysis can only proceed
after the criteria have been chosen or adopted without a choice. In
choosing a criterion of the soundness of a physical or mathematical
theory one chooses a programme for the construction of theories. In
the case of physical theories such choice is limited by the facts of
observation and experiment. But even here other requirements become
relevant, as is borne out, e.g., by the dispute between Einstein and
Bohr and their followers, not so much about the formalism of Quantum Mechanics, as about its 'intelligibility' or 'explanatory value'.[1]
In the case of mathematical theories the control by experience, if any,
is at most quite indirect; and the choice is determined more by metaphysical convictions, allegedly based on insights into the nature of
'reality', or on sound practice and tradition. These become effective
as regulative principles, i.e. as rules of conduct—the area of conduct
being the construction of mathematical theories.

[1] See, for example, the Einstein-volume of the *Library of Living Philosophers,*
ed. Schilpp, Chicago; and *Observation and Interpretation,* ed. S. Körner and
M. H. Pryce, London, 1957.

As regards the internal structure of mathematical theories there is little room for exhibition-analysis. It must be said of the rules governing statement-formation and inference within a mathematical theory, that they are either already explicitly formulated, in which case there is no need for them to be exhibited again; or that they are implicitly employed by working mathematicians, in which case it is more likely that they will eventually be brought to light by them than by philosophers who view these theories from outside rather than from within. (The axiom of choice, for example, was made explicit by the mathematician Zermelo, and the rules governing the process of substitution are now being exhibited by the mathematician Curry.)

When one comes to the general characterization of the concepts, propositions and theories of pure mathematics, and to comparison of them with other types of concepts, propositions and theories, it becomes possible for philosophical analysis, in particular exhibition-analysis, to come into its own. The philosopher is, as it were, professionally interested in comparing different disciplines and inquiries, and in ascertaining the relations between them. The contents of the last chapter of this essay are intended as a small contribution to an exhibition-analysis of pure and applied mathematics. For though it is sometimes suggested that the only subject-matter of analysis is ordinary language, and its only instrument again ordinary language, this view seems to me much too restricted. I see no reason why, say, mathematics should not be made the subject-matter of analysis or why, say, the logic of inexact concepts, in a somewhat technical presentation, should not be used as an instrument of analysis.

As regards replacement-analysis, I have been concerned with it in the first seven of these chapters. Each of the philosophies of mathematics which I discussed declares the whole or part of classical mathematics to be in some way defective, proclaims the need for replacing the defective by sound mathematical theories, and tries to meet the need by actual construction. All agree that the set-theoretical antinomies are not only obvious defects of classical mathematics, but also symptoms of deeper lying defects which they diagnose in different ways. The arguments used in the diagnosis are, as has been seen, mainly philosophical arguments, *i.e.* arguments which belong neither to natural science nor to logic.

The diagnoses—*e.g.* that a sound mathematics must be deducible from 'logical' principles, or that it must be a formalism whose consistency is proved by 'finite' methods, or that it must consist of reports on intuitive constructions, and so forth—are all philosophical diagnoses, and each leads to a programme and its implementation by a mathematical theory. If the programme is found to be unsatisfiable it is abandoned or modified. But two or more incompatible programmes may all be satisfiable and their abandonment or resurrection may be due to philosophical arguments or even fashions.

The replacement-analyses, or reconstructions, of mathematical theories in the field of 'foundations of mathematics' have thus been a joint task of mathematicians and philosophers. The defence of satisfiable programmes or of programmes not known to be unsatisfiable proceeds largely by philosophical or, to use a much abused word, metaphysical argument. The implementation or attempted implementation of a programme on the other hand is a piece of mathematics. In this essay I have sought to avoid, on the whole, any addition to the arguments for or against any programme for founding all mathematics upon one type of basic theory. I have tried rather to show the relation between philosophical programmes and their mathematical implementation. In so far as this has been done successfully, what has been given is an exhibition-analysis of philosophico-mathematical replacement-analyses.

The main aim throughout has been, on the one hand, to exhibit some general features of the reconstruction of classical mathematics in pursuance of different philosophical programmes, and on the other, to exhibit some general features of the theories of pure and applied mathematics so far constructed. The analysis may of course have failed in detail or as a whole. But if it acts as a reminder that philosophy of mathematics is neither mathematics nor mere popularization of mathematics, then it may in some degree have served the not wholly unworthy cause of opposition to the widespread retreat of philosophers from philosophy.[1]

---

[1] The general position explained and defended in sections 1–4 of this chapter derives further support from an examination of some important metamathematical theorems, in particular the Löwenheim-Skolem theorem (1920), Church's theorem (1936), the proof by Gödel (1938) of the consistency and the proof by J. P. Cohen (1963) of the independence of the continuum hypothesis. For a discussion of these matters see ' On the Relevance of Post-Gödelian Mathematics to Philosophy' in *Problems in the Philosophy of Mathematics* (ed. I. Lakatos, Amsterdam, 1967).

APPENDIX A

ON THE CLASSICAL THEORY OF REAL
NUMBERS

THE classical theory of real numbers is itself a reconstruction of a pre-classical theory implicit in the work of Newton, Leibniz and their successors. Two equivalent versions of it are due to Cantor and Dedekind respectively and variants of them are found in many modern textbooks on the theory of functions.[1] In presenting fragments of these theories here for the benefit of the non-mathematical reader I shall be following these authors. It is advisable to introduce the general reader to both theories, since, e.g., Heyting's revision of the classical theory takes Cantor's version as its starting-point, whereas Weyl's reconstruction begins with a criticism of Dedekind.

The pre-classical theory arose in Greek times from the theorem of Pythagoras. Consider an isoceles, right-angled triangle whose equal sides are of length 1 in some system of measurement. The length of the hypotenuse $x = \sqrt{1^2 + 1^2} = \sqrt{2}$. If x were rational it could be represented by a fraction p/q where p and q are of course positive integers. We may also assume that p and q *have no common divisor*. (If they have a common divisor we can always perform the division and so make the enumerator and denominator 'relatively prime'.)

From $x = \sqrt{2}$—substituting p/q for x—we get $p/q = \sqrt{2}$ and thus $p^2/q^2 = 2$ or $p^2 = 2q^2$ which means that p^2 is divisible by 2. This can be so only if p itself is divisible by 2 or even: for an odd number multiplied by an odd number, and therefore the square of an odd number, must be odd. p is therefore representable by $2r$. Substituting $2r$ for p in $p^2 = 2q^2$, we get $4r^2 = 2q^2$ or $2r^2 = q^2$. This can be so only if q^2 and thus q itself is even. But if p and q are both even, they have the common divisor 2 which is contrary to the assumption that they have no common divisor. It follows that the solution of $x^2 = 2$, *i.e.* $\sqrt{2}$ cannot

---

[1] A full treatment of Dedekind's theory is found in E. Landau's *Grundlagen der Analysis*, Leipzig, 1930; of Cantor's in H. A. Thurston's *The Number-System*, Glasgow, 1956.

be a rational number. The practice of treating $\sqrt{2}$ and other such numbers as if they obeyed all the laws obeyed by rational numbers thus needs justification.

If we perform addition, subtraction, multiplication and division on rational numbers in any order and any number of times, the result is again a rational number. But with respect to the extraction of roots (and the formation of limits of sequences) the system of rational numbers is not in this way equally 'closed'. Dedekind and Cantor therefore attempted to construct a totality of entities such that (i) it would be closed with respect to all the mentioned operations and (ii) that a subsystem of it would 'behave' in accordance with all the laws governing the rational numbers. (More precisely the subsystem would be isomorphic with the system of rational numbers.)

1. *Dedekind's reconstruction*

Landau's presentation of the theory starts with the assumption that the totality of natural numbers is given and that it is characterized by Peano's axioms, namely: (i) 1 is a natural number. (ii) To every natural number x there exists one and only one successor x'. (iii) There is no number of which the successor is 1. (iv) If $x'=y'$ then $x=y$. (v) If M is a set of natural numbers such that (a) 1 belongs to M and (b) if provided that x belongs to M, x' also belongs to M—then M comprises all the natural numbers.[1] These axioms can easily be formalized and embedded in, say, *Principia Mathematica*. The usual rules for calculating with natural numbers can be shown to hold.

Next, fractions are introduced as ordered pairs of natural numbers. Equivalence of fractions is defined, x_1/x_2 being equivalent to y_1/y_2 if and only if $x_1.y_2=y_1.x_2$. The usual rules governing calculation with fractions are established by means of definitions and theorems. Rational numbers, or more precisely *positive* rational numbers, are then introduced. A rational number is the set of all fractions which are equivalent to a fixed fraction. Thus, for example, the class $\{\frac{1}{2}, \frac{2}{4}, \frac{3}{6}, \ldots\}$ is a rational number. A rational number is called a whole number if among the fractions which it comprises (of which it is the class) there occurs $x/1$, where x is a natural number. It is shown that the whole numbers which form a subclass of the system of rational numbers have the same properties as the natural numbers, *i.e.* that the system of natural numbers is isomorphic with a system of whole numbers—a subsystem of the system of rational numbers. 'We therefore throw the natural numbers away, replace them by the corresponding whole numbers and speak henceforth (since the fractions also

[1] The principle of induction.

become superfluous) . . . only of rational numbers. (The natural numbers remain pairwise above and below the line in the notion of a fraction, and the fractions remain as the elements of the set, called rational number).'[1]

The decisive step in Dedekind's reconstruction of the old theory of real numbers is the definition of a *cut* which (in Landau's version) is intended to correspond to the naive conception of positive real numbers. A cut is a set of rational numbers such that (i) it contains some, but not all rational numbers, (ii) that every rational number belonging to the set is smaller than every rational number not belonging to it, (iii) that it contains no greatest rational number. A pictorial representation of this definition can be achieved by imagining all the positive rational numbers in their natural order marked along a straight line. Dividing this line into two parts such that the part containing the smaller rational numbers contains no greatest provides a picture of a cut. The cut is also called the 'lower class' (of the division), whereas its complement is called the 'upper class'. The members of the former are accordingly called 'lower', the members of the latter 'upper' numbers. (Cuts are designated by small Greek letters.)

Two cuts, say ξ and η, are equal if and only if every lower number of ξ is a lower number of η and vice versa; $\xi > \eta$, if and only if ξ has a lower number which is an upper number of η; $\xi < \eta$, if and only if $\eta > \xi$. It can be shown that for any two cuts ξ and η one and only one of the three relations $\xi = \eta$, $\xi > \eta$, $\xi < \eta$ must hold. Addition and multiplication of cuts are defined and shown to obey the familiar rules. (The definition of addition is arrived at as follows: (i) Let ξ and η be cuts. The set of all rational numbers of the form $X + Y$ with X lower number of ξ and Y lower number of η is shown to be a cut. (ii) It is further shown that no rational number belonging to this set can be represented as the sum of an upper number of ξ and upper number of η. (i) and (ii) having been proved the cut as constructed is called 'the sum of ξ and η' or '$\xi + \eta$'.)

It can be shown that for every rational number R, the set of all rational numbers $< R$ is a 'rational' cut; and that $=$, $>$, $<$, sum, difference, product and quotient (where it exists) of rational cuts correspond to the old concepts employed in dealing with rational numbers. 'We therefore throw the rational numbers away, replace them by the corresponding cuts and henceforth speak . . . only of cuts. (The rational numbers remain however as elements of sets used in defining the concept of a cut.)'[2] A cut which—like $\sqrt{2}$— is not rational is called irrational.

---

[1] Landau, *op. cit.*, p. 41.   [2] *Op. cit.*, p. 64.

The totality of cuts fulfils all the requirements which a suitable reconstruction of the totality of positive real numbers must fulfil. At this point Landau introduces 0 and the negative real numbers, proving the new totality consisting of the positive, negative real numbers and zero to have the required behaviour. Real numbers are written in capital Greek letters—the previous system of positive real numbers being again 'thrown away'.

We turn now to the central theorem of Dedekind's reconstruction of real numbers. Given any classification of *all* real numbers into two classes with these properties, namely that (i) there is a number in the first class and there is a number in the second class, (ii) every number of the first class is smaller than every number of the second class. Then there exists exactly one real number $\varXi$ such that every $H < \varXi$ belongs to the first and every $H > \varXi$ to the second class. The proof and the formulation of the theorem presuppose that no problem is raised by speaking of all real numbers or of an unspecified property possessed by a subclass of all real numbers and not by its complement. 'To prevent objections', Landau[1] emphasizes that in his view, '*one* number, *no* number, *two* instances, *all* things from amongst a given totality, etc., are clear word-formations. . . .' We have seen that the objections have not been prevented and that they must be taken seriously.

2. *Cantor's reconstruction of real numbers*

We assume the totality of rational numbers as given and the rules for calculating with them as given and consider sequences of rational numbers of form: $x_1, x_2, \ldots$ or briefly $\{x\}$. Of special interest among them, for our purposes, are the so-called Cauchy sequences, defined as follows:[2]

A sequence of rational numbers $x_1, x_2, \ldots$ is a Cauchy sequence, if and only if for every positive, non-zero rational number ε there exists an integer N such that $|x_p - x_q| < \varepsilon$ for $p > N$ and $q > N$. It is useful to think of x_p and x_q as points at a distance of x_p and x_q units from an origin and of $|x_p - x_q|$ as the distance between them. The definition of a Cauchy sequence is then more graphic: however small an ε we choose, there is always a member x_N in the sequence such that the distance between any two of its successors is still smaller than ε. (The operation of extracting the square root of 2 to one, two, etc. decimal places yields a Cauchy sequence of rational numbers.)

Two Cauchy sequences $\{x\}$ and $\{y\}$ are *equal* if and only if for every (positive, rational) ε there exists an integer N such that $|x_p - y_p| < \varepsilon$ for $p > N$. In other words what is required for two Cauchy

---

[1] Preface, *op. cit.*   [2] The definition is equivalent to that on p. 126.

sequences to be declared equal is that the distance between corresponding members can be as small as desired provided that we are allowed to pick a sufficiently great index for them.

The set of all Cauchy sequences which are equal to a given Cauchy sequence, say $\{x\}$, is defined as the *Cauchy number x*. (This definition is precisely like Frege's definition of an integer, or the definition of a direction as the set of all lines parallel with a given line.) It can be shown that the Cauchy numbers have all the properties which the real numbers are required to have and that they can therefore be regarded as a reconstruction of the real numbers of the 'pre-classical theory'. The relevant definitions and proofs present no difficulties. Even without going into details two features of the reconstruction are quite obvious, namely (i) the assumption that the set of all rational numbers and all its subsets are actually given, (ii) the purely existential —non-constructive—character of the definition of equality for two Cauchy numbers.

APPENDIX B

SOME SUGGESTIONS FOR FURTHER READING

THESE suggestions are limited to books in English covering the topics of the essay and readily available. Even so, many excellent texts have been omitted. Most of those mentioned contain useful bibliographies.

I *Mathematical books*

> Landau, E.: *Grundlagen der Analysis*. Translated *Foundations of Analysis* by F. Steinhardt, New York, 1957
> Courant, R., and Robbins, H.: *What is Mathematics?*, Oxford and New York, 1941
> Young, J. W. A. (editor): *Monographs on Topics of Modern Mathematics Relevant to the Elementary Field*, London, 1911; new edition, New York, 1955

The last two have been written primarily for the general reader. They give a survey of the main topics occupying contemporary working mathematicians and give a fair idea of their style of reasoning.

II *General works on the foundations of mathematics*

> Black, M.: *The Nature of Mathematics*, London, 1933
> Wilder, R. L.: *Introduction to the Foundations of Mathematics*, New York, 1952
> Fraenkel, A. A., Bar-Hillel, Y.: *Foundations of Set Theory*, Amsterdam, 1958

The first of these books gives more attention to philosophical questions than the other two. The last contains a thorough survey of the present state of set-theory and reviews the many formalisms used by mathematical logicians.

III *Books of mainly logicist tendency*

> Frege, G.: *Die Grundlagen der Arithmetik*. German text and English translation by J. L. Austin, Oxford, 1950

Frege, G.: *Translations from the Philosophical Writings of Frege* by P. Geach and M. Black, Oxford, 1952

Russell, B.: *Introduction to Mathematical Philosophy*, 2nd ed., London, 1938

Quine, W. V.: *From a Logical Point of View*, Cambridge, Mass., 1953. (Contains 'New Foundations for Mathematical Logic')

Quine, W. V.: *Mathematical Logic*, revised edition, Cambridge, Mass., 1955

Church, A.: *Introduction to Mathematical Logic*, vol. I., Princeton, 1956

The last two are important recent treatises.

IV *Books of mainly formalist tendency*

Hilbert, D. and Ackermann, W.: *Grundzüge der Theoretischen Logik*, 3rd ed. Translated as *Principles of Mathematical Logic* by L. M. Hammond, G. L. Leckie, F. Steinhardt, edited by R. E. Luce, New York, 1950

Curry, H. B.: *Outlines of a Formalist Philosophy of Mathematics*, Amsterdam, 1951

Kleene, S. C.: *Introduction to Metamathematics*, Amsterdam, 1952

The last is an important recent treatise. The second expounds and defends a formalist philosophy of mathematics.

V *Books of intuitionist tendency*

Heyting, A.: *Intuitionism—An Introduction*, Amsterdam, 1956

The only text comprehensively introductory in English.

VI *Other works*

Mostowski, A.: *Sentences Undecidable in Formalized Arithmetic*, Amsterdam, 1952

Mostowski, A.: *Thirty Years of Foundational Studies* (1930–1964), Oxford, 1966

Tarski, A.: *Introduction to Logic and the Methodology of Deductive Sciences*, 2nd ed., London, 1946

The last is one of the best elementary introductions to modern logic.

INDEX

(italicized page numbers indicate definitions or explanations of terms)

195

A CATALOG OF SELECTED

DOVER BOOKS

IN ALL FIELDS OF INTEREST

DRAWINGS OF REMBRANDT, edited by Seymour Slive. Updated Lippmann, Hofstede de Groot edition, with definitive scholarly apparatus. All portraits, biblical sketches, landscapes, nudes. Oriental figures, classical studies, together with selection of work by followers. 550 illustrations. Total of 630pp. 9⅛ × 12¼.
21485-0, 21486-9 Pa., Two-vol. set $29.90

GHOST AND HORROR STORIES OF AMBROSE BIERCE, Ambrose Bierce. 24 tales vividly imagined, strangely prophetic, and decades ahead of their time in technical skill: "The Damned Thing," "An Inhabitant of Carcosa," "The Eyes of the Panther," "Moxon's Master," and 20 more. 199pp. 5⅜ × 8½. 20767-6 Pa. $4.95

ETHICAL WRITINGS OF MAIMONIDES, Maimonides. Most significant ethical works of great medieval sage, newly translated for utmost precision, readability. Laws Concerning Character Traits, Eight Chapters, more. 192pp. 5⅜ × 8½.
24522-5 Pa. $4.50

THE EXPLORATION OF THE COLORADO RIVER AND ITS CANYONS, J. W. Powell. Full text of Powell's 1,000-mile expedition down the fabled Colorado in 1869. Superb account of terrain, geology, vegetation, Indians, famine, mutiny, treacherous rapids, mighty canyons, during exploration of last unknown part of continental U.S. 400pp. 5⅜ × 8½. 20094-9 Pa. $7.95

HISTORY OF PHILOSOPHY, Julián Marías. Clearest one-volume history on the market. Every major philosopher and dozens of others, to Existentialism and later. 505pp. 5⅜ × 8½. 21739-6 Pa. $9.95

ALL ABOUT LIGHTNING, Martin A. Uman. Highly readable nontechnical survey of nature and causes of lightning, thunderstorms, ball lightning, St. Elmo's Fire, much more. Illustrated. 192pp. 5⅜ × 8½. 25237-X Pa. $5.95

SAILING ALONE AROUND THE WORLD, Captain Joshua Slocum. First man to sail around the world, alone, in small boat. One of great feats of seamanship told in delightful manner. 67 illustrations. 294pp. 5⅜ × 8½. 20326-3 Pa. $4.95

LETTERS AND NOTES ON THE MANNERS, CUSTOMS AND CONDITIONS OF THE NORTH AMERICAN INDIANS, George Catlin. Classic account of life among Plains Indians: ceremonies, hunt, warfare, etc. 312 plates. 572pp. of text. 6⅛ × 9¼. 22118-0, 22119-9, Pa., Two-vol. set $17.90

ALASKA: The Harriman Expedition, 1899, John Burroughs, John Muir, et al. Informative, engrossing accounts of two-month, 9,000-mile expedition. Native peoples, wildlife, forests, geography, salmon industry, glaciers, more. Profusely illustrated. 240 black-and-white line drawings. 124 black-and-white photographs. 3 maps. Index. 576pp. 5⅜ × 8½. 25109-8 Pa. $11.95

THE BOOK OF BEASTS: Being a Translation from a Latin Bestiary of the Twelfth Century, T. H. White. Wonderful catalog of real and fanciful beasts: manticore, griffin, phoenix, amphivius, jaculus, many more. White's witty erudite commentary on scientific, historical aspects enhances fascinating glimpse of medieval mind. Illustrated. 296pp. 5⅜ × 8¼. (Available in U.S. only)   24609-4 Pa. $6.95

FRANK LLOYD WRIGHT: Architecture and Nature with 160 Illustrations, Donald Hoffmann. Profusely illustrated study of influence of nature—especially prairie—on Wright's designs for Fallingwater, Robie House, Guggenheim Museum, other masterpieces. 96pp. 9¼ × 10¾.   25098-9 Pa. $8.95

FRANK LLOYD WRIGHT'S FALLINGWATER, Donald Hoffmann. Wright's famous waterfall house: planning and construction of organic idea. History of site, owners, Wright's personal involvement. Photographs of various stages of building. Preface by Edgar Kaufmann, Jr. 100 illustrations. 112pp. 9¼ × 10.
23671-4 Pa. $8.95

YEARS WITH FRANK LLOYD WRIGHT: Apprentice to Genius, Edgar Tafel. Insightful memoir by a former apprentice presents a revealing portrait of Wright the man, the inspired teacher, the greatest American architect. 372 black-and-white illustrations. Preface. Index. vi + 228pp. 8¼ × 11.   24801-1 Pa. $10.95

THE STORY OF KING ARTHUR AND HIS KNIGHTS, Howard Pyle. Enchanting version of King Arthur fable has delighted generations with imaginative narratives of exciting adventures and unforgettable illustrations by the author. 41 illustrations. xviii + 313pp. 6⅛ × 9¼.   21445-1 Pa. $6.95

THE GODS OF THE EGYPTIANS, E. A. Wallis Budge. Thorough coverage of numerous gods of ancient Egypt by foremost Egyptologist. Information on evolution of cults, rites and gods; the cult of Osiris; the Book of the Dead and its rites; the sacred animals and birds; Heaven and Hell; and more. 956pp. 6⅛ × 9¼.
22055-9, 22056-7 Pa., Two-vol. set $21.90

A THEOLOGICO-POLITICAL TREATISE, Benedict Spinoza. Also contains unfinished *Political Treatise*. Great classic on religious liberty, theory of government on common consent. R. Elwes translation. Total of 421pp. 5⅜ × 8½.
20249-6 Pa. $7.95

INCIDENTS OF TRAVEL IN CENTRAL AMERICA, CHIAPAS, AND YUCATAN, John L. Stephens. Almost single-handed discovery of Maya culture; exploration of ruined cities, monuments, temples; customs of Indians. 115 drawings. 892pp. 5⅜ × 8½.   22404-X, 22405-8 Pa., Two-vol. set $17.90

LOS CAPRICHOS, Francisco Goya. 80 plates of wild, grotesque monsters and caricatures. Prado manuscript included. 183pp. 6⅜ × 9⅜.   22384-1 Pa. $5.95

AUTOBIOGRAPHY: The Story of My Experiments with Truth, Mohandas K. Gandhi. Not hagiography, but Gandhi in his own words. Boyhood, legal studies, purification, the growth of the Satyagraha (nonviolent protest) movement. Critical, inspiring work of the man who freed India. 480pp. 5⅜ × 8½. (Available in U.S. only)
24593-4 Pa. $6.95

ILLUSTRATED DICTIONARY OF HISTORIC ARCHITECTURE, edited by Cyril M. Harris. Extraordinary compendium of clear, concise definitions for over 5,000 important architectural terms complemented by over 2,000 line drawings. Covers full spectrum of architecture from ancient ruins to 20th-century Modernism. Preface. 592pp. 7½ × 9⅝. 24444-X Pa. $15.95

THE NIGHT BEFORE CHRISTMAS, Clement C. Moore. Full text, and woodcuts from original 1848 book. Also critical, historical material. 19 illustrations. 40pp. 4⅝ × 6. 22797-9 Pa. $2.50

THE LESSON OF JAPANESE ARCHITECTURE: 165 Photographs, Jiro Harada. Memorable gallery of 165 photographs taken in the 1930s of exquisite Japanese homes of the well-to-do and historic buildings. 13 line diagrams. 192pp. 8⅜ × 11¼. 24778-3 Pa. $10.95

THE AUTOBIOGRAPHY OF CHARLES DARWIN AND SELECTED LETTERS, edited by Francis Darwin. The fascinating life of eccentric genius composed of an intimate memoir by Darwin (intended for his children); commentary by his son, Francis; hundreds of fragments from notebooks, journals, papers; and letters to and from Lyell, Hooker, Huxley, Wallace and Henslow. xi + 365pp. 5⅜ × 8. 20479-0 Pa. $6.95

WONDERS OF THE SKY: Observing Rainbows, Comets, Eclipses, the Stars and Other Phenomena, Fred Schaaf. Charming, easy-to-read poetic guide to all manner of celestial events visible to the naked eye. Mock suns, glories, Belt of Venus, more. Illustrated. 299pp. 5¼ × 8¼. 24402-4 Pa. $8.95

BURNHAM'S CELESTIAL HANDBOOK, Robert Burnham, Jr. Thorough guide to the stars beyond our solar system. Exhaustive treatment. Alphabetical by constellation: Andromeda to Cetus in Vol. 1; Chamaeleon to Orion in Vol. 2; and Pavo to Vulpecula in Vol. 3. Hundreds of illustrations. Index in Vol. 3. 2,000pp. 6⅛ × 9¼. 23567-X, 23568-8, 23673-0 Pa., Three-vol. set $41.85

STAR NAMES: Their Lore and Meaning, Richard Hinckley Allen. Fascinating history of names various cultures have given to constellations and literary and folkloristic uses that have been made of stars. Indexes to subjects. Arabic and Greek names. Biblical references. Bibliography. 563pp. 5⅜ × 8½. 21079-0 Pa. $8.95

THIRTY YEARS THAT SHOOK PHYSICS: The Story of Quantum Theory, George Gamow. Lucid, accessible introduction to influential theory of energy and matter. Careful explanations of Dirac's anti-particles, Bohr's model of the atom, much more. 12 plates. Numerous drawings. 240pp. 5⅜ × 8½. 24895-X Pa. $6.95

CHINESE DOMESTIC FURNITURE IN PHOTOGRAPHS AND MEASURED DRAWINGS, Gustav Ecke. A rare volume, now affordably priced for antique collectors, furniture buffs and art historians. Detailed review of styles ranging from early Shang to late Ming. Unabridged republication. 161 black-and-white drawings, photos. Total of 224pp. 8⅜ × 11¼. (Available in U.S. only) 25171-3 Pa. $14.95

VINCENT VAN GOGH: A Biography, Julius Meier-Graefe. Dynamic, penetrating study of artist's life, relationship with brother, Theo, painting techniques, travels, more. Readable, engrossing. 160pp. 5⅜ × 8½. (Available in U.S. only) 25253-1 Pa. $4.95

HOW TO WRITE, Gertrude Stein. Gertrude Stein claimed anyone could understand her unconventional writing—here are clues to help. Fascinating improvisations, language experiments, explanations illuminate Stein's craft and the art of writing. Total of 414pp. 4⅝ × 6⅜. 23144-5 Pa. $6.95

ADVENTURES AT SEA IN THE GREAT AGE OF SAIL: Five Firsthand Narratives, edited by Elliot Snow. Rare true accounts of exploration, whaling, shipwreck, fierce natives, trade, shipboard life, more. 33 illustrations. Introduction. 353pp. 5⅜ × 8½. 25177-2 Pa. $9.95

THE HERBAL OR GENERAL HISTORY OF PLANTS, John Gerard. Classic descriptions of about 2,850 plants—with over 2,700 illustrations—includes Latin and English names, physical descriptions, varieties, time and place of growth, more. 2,706 illustrations. xlv + 1,678pp. 8½ × 12¼. 23147-X Cloth. $75.00

DOROTHY AND THE WIZARD IN OZ, L. Frank Baum. Dorothy and the Wizard visit the center of the Earth, where people are vegetables, glass houses grow and Oz characters reappear. Classic sequel to Wizard of Oz. 256pp. 5⅜ × 8. 24714-7 Pa. $5.95

SONGS OF EXPERIENCE: Facsimile Reproduction with 26 Plates in Full Color, William Blake. This facsimile of Blake's original "Illuminated Book" reproduces 26 full-color plates from a rare 1826 edition. Includes "The Tyger," "London," "Holy Thursday," and other immortal poems. 26 color plates. Printed text of poems. 48pp. 5¼ × 7. 24636-1 Pa. $3.95

SONGS OF INNOCENCE, William Blake. The first and most popular of Blake's famous "Illuminated Books," in a facsimile edition reproducing all 31 brightly colored plates. Additional printed text of each poem. 64pp. 5¼ × 7. 22764-2 Pa. $3.95

PRECIOUS STONES, Max Bauer. Classic, thorough study of diamonds, rubies, emeralds, garnets, etc.: physical character, occurrence, properties, use, similar topics. 20 plates, 8 in color. 94 figures. 659pp. 6⅛ × 9¼. 21910-0, 21911-9 Pa., Two-vol. set $15.90

ENCYCLOPEDIA OF VICTORIAN NEEDLEWORK, S. F. A. Caulfeild and Blanche Saward. Full, precise descriptions of stitches, techniques for dozens of needlecrafts—most exhaustive reference of its kind. Over 800 figures. Total of 679pp. 8⅛ × 11. 22800-2, 22801-0 Pa., Two-vol. set $23.90

THE MARVELOUS LAND OF OZ, L. Frank Baum. Second Oz book, the Scarecrow and Tin Woodman are back with hero named Tip, Oz magic. 136 illustrations. 287pp. 5⅜ × 8½. 20692-0 Pa. $5.95

WILD FOWL DECOYS, Joel Barber. Basic book on the subject, by foremost authority and collector. Reveals history of decoy making and rigging, place in American culture, different kinds of decoys, how to make them, and how to use them. 140 plates. 156pp. 7⅞ × 10¾. 20011-6 Pa. $8.95

HISTORY OF LACE, Mrs. Bury Palliser. Definitive, profusely illustrated chronicle of lace from earliest times to late 19th century. Laces of Italy, Greece, England, France, Belgium, etc. Landmark of needlework scholarship. 266 illustrations. 672pp. 6⅛ × 9¼. 24742-2 Pa. $16.95

ILLUSTRATED GUIDE TO SHAKER FURNITURE, Robert Meader. All furniture and appurtenances, with much on unknown local styles. 235 photos. 146pp. 9 × 12.   22819-3 Pa. $8.95

WHALE SHIPS AND WHALING: A Pictorial Survey, George Francis Dow. Over 200 vintage engravings, drawings, photographs of barks, brigs, cutters, other vessels. Also harpoons, lances, whaling guns, many other artifacts. Comprehensive text by foremost authority. 207 black-and-white illustrations. 288pp. 6 × 9.
                                                              24808-9 Pa. $9.95

THE BERTRAMS, Anthony Trollope. Powerful portrayal of blind self-will and thwarted ambition includes one of Trollope's most heartrending love stories. 497pp. 5⅜ × 8½.   25119-5 Pa. $9.95

ADVENTURES WITH A HAND LENS, Richard Headstrom. Clearly written guide to observing and studying flowers and grasses, fish scales, moth and insect wings, egg cases, buds, feathers, seeds, leaf scars, moss, molds, ferns, common crystals, etc.—all with an ordinary, inexpensive magnifying glass. 209 exact line drawings aid in your discoveries. 220pp. 5⅜ × 8½.   23330-8 Pa. $5.95

RODIN ON ART AND ARTISTS, Auguste Rodin. Great sculptor's candid, wide-ranging comments on meaning of art; great artists; relation of sculpture to poetry, painting, music; philosophy of life, more. 76 superb black-and-white illustrations of Rodin's sculpture, drawings and prints. 119pp. 8⅜ × 11¼.   24487-3 Pa. $7.95

FIFTY CLASSIC FRENCH FILMS, 1912–1982: A Pictorial Record, Anthony Slide. Memorable stills from Grand Illusion, Beauty and the Beast, Hiroshima, Mon Amour, many more. Credits, plot synopses, reviews, etc. 160pp. 8¼ × 11.
                                                              25256-6 Pa. $11.95

THE PRINCIPLES OF PSYCHOLOGY, William James. Famous long course complete, unabridged. Stream of thought, time perception, memory, experimental methods; great work decades ahead of its time. 94 figures. 1,391pp. 5⅜ × 8½.
                                        20381-6, 20382-4 Pa., Two-vol. set $25.90

BODIES IN A BOOKSHOP, R. T. Campbell. Challenging mystery of blackmail and murder with ingenious plot and superbly drawn characters. In the best tradition of British suspense fiction. 192pp. 5⅜ × 8½.   24720-1 Pa. $4.95

CALLAS: Portrait of a Prima Donna, George Jellinek. Renowned commentator on the musical scene chronicles incredible career and life of the most controversial, fascinating, influential operatic personality of our time. 64 black-and-white photographs. 416pp. 5⅜ × 8¼.   25047-4 Pa. $8.95

GEOMETRY, RELATIVITY AND THE FOURTH DIMENSION, Rudolph Rucker. Exposition of fourth dimension, concepts of relativity as Flatland characters continue adventures. Popular, easily followed yet accurate, profound. 141 illustrations. 133pp. 5⅜ × 8½.   23400-2 Pa. $4.95

HOUSEHOLD STORIES BY THE BROTHERS GRIMM, with pictures by Walter Crane. 53 classic stories—Rumpelstiltskin, Rapunzel, Hansel and Gretel, the Fisherman and his Wife, Snow White, Tom Thumb, Sleeping Beauty, Cinderella, and so much more—lavishly illustrated with original 19th-century drawings. 114 illustrations. x + 269pp. 5⅜ × 8½.   21080-4 Pa. $4.95

SUNDIALS, Albert Waugh. Far and away the best, most thorough coverage of ideas, mathematics concerned, types, construction, adjusting anywhere. Over 100 illustrations. 230pp. 5⅜ × 8½. 22947-5 Pa. $5.95

PICTURE HISTORY OF THE NORMANDIE: With 190 Illustrations, Frank O. Braynard. Full story of legendary French ocean liner: Art Deco interiors, design innovations, furnishings, celebrities, maiden voyage, tragic fire, much more. Extensive text. 144pp. 8⅜ × 11¾. 25257-4 Pa. $10.95

THE FIRST AMERICAN COOKBOOK: A Facsimile of "American Cookery," 1796, Amelia Simmons. Facsimile of the first American-written cookbook published in the United States contains authentic recipes for colonial favorites—pumpkin pudding, winter squash pudding, spruce beer, Indian slapjacks, and more. Introductory Essay and Glossary of colonial cooking terms. 80pp. 5⅜ × 8½. 24710-4 Pa. $3.50

101 PUZZLES IN THOUGHT AND LOGIC, C. R. Wylie, Jr. Solve murders and robberies, find out which fishermen are liars, how a blind man could possibly identify a color—purely by your own reasoning! 107pp. 5⅜ × 8½. 20367-0 Pa. $2.95

ANCIENT EGYPTIAN MYTHS AND LEGENDS, Lewis Spence. Examines animism, totemism, fetishism, creation myths, deities, alchemy, art and magic, other topics. Over 50 illustrations. 432pp. 5⅜ × 8½. 26525-0 Pa. $8.95

ANTHROPOLOGY AND MODERN LIFE, Franz Boas. Great anthropologist's classic treatise on race and culture. Introduction by Ruth Bunzel. Only inexpensive paperback edition. 255pp. 5⅜ × 8½. 25245-0 Pa. $6.95

THE TALE OF PETER RABBIT, Beatrix Potter. The inimitable Peter's terrifying adventure in Mr. McGregor's garden, with all 27 wonderful, full-color Potter illustrations. 55pp. 4¼ × 5½. (Available in U.S. only) 22827-4 Pa. $1.75

THREE PROPHETIC SCIENCE FICTION NOVELS, H. G. Wells. *When the Sleeper Wakes, A Story of the Days to Come* and *The Time Machine* (full version). 335pp. 5⅜ × 8½. (Available in U.S. only) 20605-X Pa. $8.95

APICIUS COOKERY AND DINING IN IMPERIAL ROME, edited and translated by Joseph Dommers Vehling. Oldest known cookbook in existence offers readers a clear picture of what foods Romans ate, how they prepared them, etc. 49 illustrations. 301pp. 6⅛ × 9¼. 23563-7 Pa. $7.95

SHAKESPEARE LEXICON AND QUOTATION DICTIONARY, Alexander Schmidt. Full definitions, locations, shades of meaning of every word in plays and poems. More than 50,000 exact quotations. 1,485pp. 6½ × 9¼. 22726-X, 22727-8 Pa., Two-vol. set $31.90

THE WORLD'S GREAT SPEECHES, edited by Lewis Copeland and Lawrence W. Lamm. Vast collection of 278 speeches from Greeks to 1970. Powerful and effective models; unique look at history. 842pp. 5⅜ × 8½. 20468-5 Pa. $12.95

THE BLUE FAIRY BOOK, Andrew Lang. The first, most famous collection, with many familiar tales: Little Red Riding Hood, Aladdin and the Wonderful Lamp, Puss in Boots, Sleeping Beauty, Hansel and Gretel, Rumpelstiltskin; 37 in all. 138 illustrations. 390pp. 5⅜ × 8½. 21437-0 Pa. $6.95

THE STORY OF THE CHAMPIONS OF THE ROUND TABLE, Howard Pyle. Sir Launcelot, Sir Tristram and Sir Percival in spirited adventures of love and triumph retold in Pyle's inimitable style. 50 drawings, 31 full-page. xviii + 329pp. 6½ × 9¼. 21883-X Pa. $7.95

THE MYTHS OF THE NORTH AMERICAN INDIANS, Lewis Spence. Myths and legends of the Algonquins, Iroquois, Pawnees and Sioux with comprehensive historical and ethnological commentary. 36 illustrations. 5⅜ × 8½.
25967-6 Pa. $8.95

GREAT DINOSAUR HUNTERS AND THEIR DISCOVERIES, Edwin H. Colbert. Fascinating, lavishly illustrated chronicle of dinosaur research, 1820s to 1960. Achievements of Cope, Marsh, Brown, Buckland, Mantell, Huxley, many others. 384pp. 5¼ × 8¼. 24701-5 Pa. $7.95

THE TASTEMAKERS, Russell Lynes. Informal, illustrated social history of American taste 1850s–1950s. First popularized categories Highbrow, Lowbrow, Middlebrow. 129 illustrations. New (1979) afterword. 384pp. 6 × 9.
23993-4 Pa. $8.95

DOUBLE CROSS PURPOSES, Ronald A. Knox. A treasure hunt in the Scottish Highlands, an old map, unidentified corpse, surprise discoveries keep reader guessing in this cleverly intricate tale of financial skullduggery. 2 black-and-white maps. 320pp. 5⅜ × 8½. (Available in U.S. only) 25032-6 Pa. $6.95

AUTHENTIC VICTORIAN DECORATION AND ORNAMENTATION IN FULL COLOR: 46 Plates from "Studies in Design," Christopher Dresser. Superb full-color lithographs reproduced from rare original portfolio of a major Victorian designer. 48pp. 9¼ × 12¼. 25083-0 Pa. $7.95

PRIMITIVE ART, Franz Boas. Remains the best text ever prepared on subject, thoroughly discussing Indian, African, Asian, Australian, and, especially, Northern American primitive art. Over 950 illustrations show ceramics, masks, totem poles, weapons, textiles, paintings, much more. 376pp. 5⅜ × 8. 20025-6 Pa. $7.95

SIDELIGHTS ON RELATIVITY, Albert Einstein. Unabridged republication of two lectures delivered by the great physicist in 1920–21. *Ether and Relativity* and *Geometry and Experience*. Elegant ideas in nonmathematical form, accessible to intelligent layman. vi + 56pp. 5⅜ × 8½. 24511-X Pa. $3.95

THE WIT AND HUMOR OF OSCAR WILDE, edited by Alvin Redman. More than 1,000 ripostes, paradoxes, wisecracks: Work is the curse of the drinking classes, I can resist everything except temptation, etc. 258pp. 5⅜ × 8½. 20602-5 Pa. $4.95

ADVENTURES WITH A MICROSCOPE, Richard Headstrom. 59 adventures with clothing fibers, protozoa, ferns and lichens, roots and leaves, much more. 142 illustrations. 232pp. 5⅜ × 8½. 23471-1 Pa. $3.95

PLANTS OF THE BIBLE, Harold N. Moldenke and Alma L. Moldenke. Standard reference to all 230 plants mentioned in Scriptures. Latin name, biblical reference, uses, modern identity, much more. Unsurpassed encyclopedic resource for scholars, botanists, nature lovers, students of Bible. Bibliography. Indexes. 123 black-and-white illustrations. 384pp. 6 × 9. 25069-5 Pa. $8.95

FAMOUS AMERICAN WOMEN: A Biographical Dictionary from Colonial Times to the Present, Robert McHenry, ed. From Pocahontas to Rosa Parks, 1,035 distinguished American women documented in separate biographical entries. Accurate, up-to-date data, numerous categories, spans 400 years. Indices. 493pp. 6½ × 9¼. 24523-3 Pa. $10.95

THE FABULOUS INTERIORS OF THE GREAT OCEAN LINERS IN HISTORIC PHOTOGRAPHS, William H. Miller, Jr. Some 200 superb photographs capture exquisite interiors of world's great "floating palaces"—1890s to 1980s: *Titanic, Ile de France, Queen Elizabeth, United States, Europa,* more. Approx. 200 black-and-white photographs. Captions. Text. Introduction. 160pp. 8⅞ × 11¾. 24756-2 Pa. $9.95

THE GREAT LUXURY LINERS, 1927–1954: A Photographic Record, William H. Miller, Jr. Nostalgic tribute to heyday of ocean liners. 186 photos of *Ile de France, Normandie, Leviathan, Queen Elizabeth, United States,* many others. Interior and exterior views. Introduction. Captions. 160pp. 9 × 12. 24056-8 Pa. $10.95

A NATURAL HISTORY OF THE DUCKS, John Charles Phillips. Great landmark of ornithology offers complete detailed coverage of nearly 200 species and subspecies of ducks: gadwall, sheldrake, merganser, pintail, many more. 74 full-color plates, 102 black-and-white. Bibliography. Total of 1,920pp. 8⅝ × 11¼. 25141-1, 25142-X Cloth., Two-vol. set $100.00

THE SEAWEED HANDBOOK: An Illustrated Guide to Seaweeds from North Carolina to Canada, Thomas F. Lee. Concise reference covers 78 species. Scientific and common names, habitat, distribution, more. Finding keys for easy identification. 224pp. 5⅜ × 8½. 25215-9 Pa. $6.95

THE TEN BOOKS OF ARCHITECTURE: The 1755 Leoni Edition, Leon Battista Alberti. Rare classic helped introduce the glories of ancient architecture to the Renaissance. 68 black-and-white plates. 336pp. 8⅜ × 11¼. 25239-6 Pa. $14.95

MISS MACKENZIE, Anthony Trollope. Minor masterpieces by Victorian master unmasks many truths about life in 19th-century England. First inexpensive edition in years. 392pp. 5⅜ × 8½. 25201-9 Pa. $8.95

THE RIME OF THE ANCIENT MARINER, Gustave Doré, Samuel Taylor Coleridge. Dramatic engravings considered by many to be his greatest work. The terrifying space of the open sea, the storms and whirlpools of an unknown ocean, the ice of Antarctica, more—all rendered in a powerful, chilling manner. Full text. 38 plates. 77pp. 9¼ × 12. 22305-1 Pa. $4.95

THE EXPEDITIONS OF ZEBULON MONTGOMERY PIKE, Zebulon Montgomery Pike. Fascinating firsthand accounts (1805–6) of exploration of Mississippi River, Indian wars, capture by Spanish dragoons, much more. 1,088pp. 5⅜ × 8½. 25254-X, 25255-8 Pa., Two-vol. set $25.90

A CONCISE HISTORY OF PHOTOGRAPHY: Third Revised Edition, Helmut Gernsheim. Best one-volume history—camera obscura, photochemistry, daguerreotypes, evolution of cameras, film, more. Also artistic aspects—landscape, portraits, fine art, etc. 281 black-and-white photographs. 26 in color. 176pp. 8⅜ × 11¼.
25128-4 Pa. $14.95

THE DORÉ BIBLE ILLUSTRATIONS, Gustave Doré. 241 detailed plates from the Bible: the Creation scenes, Adam and Eve, Flood, Babylon, battle sequences, life of Jesus, etc. Each plate is accompanied by the verses from the King James version of the Bible. 241pp. 9 × 12. 23004-X Pa. $9.95

WANDERINGS IN WEST AFRICA, Richard F. Burton. Great Victorian scholar/adventurer's invaluable descriptions of African tribal rituals, fetishism, culture, art, much more. Fascinating 19th-century account. 624pp. 5⅜ × 8½. 26890-X Pa. $12.95

FLATLAND, E. A. Abbott. Intriguing and enormously popular science-fiction classic explores the complexities of trying to survive as a two-dimensional being in a three-dimensional world. Amusingly illustrated by the author. 16 illustrations. 103pp. 5⅜ × 8½. 20001-9 Pa. $2.50

THE HISTORY OF THE LEWIS AND CLARK EXPEDITION, Meriwether Lewis and William Clark, edited by Elliott Coues. Classic edition of Lewis and Clark's day-by-day journals that later became the basis for U.S. claims to Oregon and the West. Accurate and invaluable geographical, botanical, biological, meteorological and anthropological material. Total of 1,508pp. 5⅜ × 8½.
21268-8, 21269-6, 21270-X Pa., Three-vol. set $29.85

LANGUAGE, TRUTH AND LOGIC, Alfred J. Ayer. Famous, clear introduction to Vienna, Cambridge schools of Logical Positivism. Role of philosophy, elimination of metaphysics, nature of analysis, etc. 160pp. 5⅜ × 8½. (Available in U.S. and Canada only) 20010-8 Pa. $3.95

MATHEMATICS FOR THE NONMATHEMATICIAN, Morris Kline. Detailed, college-level treatment of mathematics in cultural and historical context, with numerous exercises. For liberal arts students. Preface. Recommended Reading Lists. Tables. Index. Numerous black-and-white figures. xvi + 641pp. 5⅜ × 8½.
24823-2 Pa. $11.95

HANDBOOK OF PICTORIAL SYMBOLS, Rudolph Modley. 3,250 signs and symbols, many systems in full; official or heavy commercial use. Arranged by subject. Most in Pictorial Archive series. 143pp. 8⅛ × 11. 23357-X Pa. $7.95

INCIDENTS OF TRAVEL IN YUCATAN, John L. Stephens. Classic (1843) exploration of jungles of Yucatan, looking for evidences of Maya civilization. Travel adventures, Mexican and Indian culture, etc. Total of 669pp. 5⅜ × 8½.
20926-1, 20927-X Pa., Two-vol. set $11.90

CATALOG OF DOVER BOOKS

DEGAS: An Intimate Portrait, Ambroise Vollard. Charming, anecdotal memoir by famous art dealer of one of the greatest 19th-century French painters. 14 black-and-white illustrations. Introduction by Harold L. Van Doren. 96pp. 5⅜ × 8½.
25131-4 Pa. $4.95

PERSONAL NARRATIVE OF A PILGRIMAGE TO AL-MADINAH AND MECCAH, Richard F. Burton. Great travel classic by remarkably colorful personality. Burton, disguised as a Moroccan, visited sacred shrines of Islam, narrowly escaping death. 47 illustrations. 959pp. 5⅜ × 8½.
21217-3, 21218-1 Pa., Two-vol. set $19.90

PHRASE AND WORD ORIGINS, A. H. Holt. Entertaining, reliable, modern study of more than 1,200 colorful words, phrases, origins and histories. Much unexpected information. 254pp. 5⅜ × 8½.
20758-7 Pa. $5.95

THE RED THUMB MARK, R. Austin Freeman. In this first Dr. Thorndyke case, the great scientific detective draws fascinating conclusions from the nature of a single fingerprint. Exciting story, authentic science. 320pp. 5⅜ × 8½. (Available in U.S. only)
25210-8 Pa. $6.95

AN EGYPTIAN HIEROGLYPHIC DICTIONARY, E. A. Wallis Budge. Monumental work containing about 25,000 words or terms that occur in texts ranging from 3000 B.C. to 600 A.D. Each entry consists of a transliteration of the word, the word in hieroglyphs, and the meaning in English. 1,314pp. 6⅝ × 10.
23615-3, 23616-1 Pa., Two-vol. set $35.90

THE COMPLEAT STRATEGYST: Being a Primer on the Theory of Games of Strategy, J. D. Williams. Highly entertaining classic describes, with many illustrated examples, how to select best strategies in conflict situations. Prefaces. Appendices. xvi + 268pp. 5⅜ × 8½.
25101-2 Pa. $6.95

THE ROAD TO OZ, L. Frank Baum. Dorothy meets the Shaggy Man, little Button-Bright and the Rainbow's beautiful daughter in this delightful trip to the magical Land of Oz. 272pp. 5⅜ × 8.
25208-6 Pa. $5.95

POINT AND LINE TO PLANE, Wassily Kandinsky. Seminal exposition of role of point, line, other elements in nonobjective painting. Essential to understanding 20th-century art. 127 illustrations. 192pp. 6½ × 9¼.
23808-3 Pa. $5.95

LADY ANNA, Anthony Trollope. Moving chronicle of Countess Lovel's bitter struggle to win for herself and daughter Anna their rightful rank and fortune—perhaps at cost of sanity itself. 384pp. 5⅜ × 8½.
24669-8 Pa. $8.95

EGYPTIAN MAGIC, E. A. Wallis Budge. Sums up all that is known about magic in Ancient Egypt: the role of magic in controlling the gods, powerful amulets that warded off evil spirits, scarabs of immortality, use of wax images, formulas and spells, the secret name, much more. 253pp. 5⅜ × 8½. 22681-6 Pa. $4.50

THE DANCE OF SIVA, Ananda Coomaraswamy. Preeminent authority unfolds the vast metaphysic of India: the revelation of her art, conception of the universe, social organization, etc. 27 reproductions of art masterpieces. 192pp. 5⅜ × 8½.
24817-8 Pa. $6.95

CHRISTMAS CUSTOMS AND TRADITIONS, Clement A. Miles. Origin, evolution, significance of religious, secular practices. Caroling, gifts, yule logs, much more. Full, scholarly yet fascinating; non-sectarian. 400pp. 5⅜ × 8½.
23354-5 Pa. $6.95

THE HUMAN FIGURE IN MOTION, Eadweard Muybridge. More than 4,500 stopped-action photos, in action series, showing undraped men, women, children jumping, lying down, throwing, sitting, wrestling, carrying, etc. 390pp. 7⅞ × 10⅝.
20204-6 Cloth. $24.95

THE MAN WHO WAS THURSDAY, Gilbert Keith Chesterton. Witty, fast-paced novel about a club of anarchists in turn-of-the-century London. Brilliant social, religious, philosophical speculations. 128pp. 5⅜ × 8½.
25121-7 Pa. $3.95

A CÉZANNE SKETCHBOOK: Figures, Portraits, Landscapes and Still Lifes, Paul Cézanne. Great artist experiments with tonal effects, light, mass, other qualities in over 100 drawings. A revealing view of developing master painter, precursor of Cubism. 102 black-and-white illustrations. 144pp. 8¾ × 6⅝.
24790-2 Pa. $6.95

AN ENCYCLOPEDIA OF BATTLES: Accounts of Over 1,560 Battles from 1479 B.C. to the Present, David Eggenberger. Presents essential details of every major battle in recorded history, from the first battle of Megiddo in 1479 B.C. to Grenada in 1984. List of Battle Maps. New Appendix covering the years 1967–1984. Index. 99 illustrations. 544pp. 6½ × 9¼.
24913-1 Pa. $14.95

AN ETYMOLOGICAL DICTIONARY OF MODERN ENGLISH, Ernest Weekley. Richest, fullest work, by foremost British lexicographer. Detailed word histories. Inexhaustible. Total of 856pp. 6½ × 9¼.
21873-2, 21874-0 Pa., Two-vol. set $19.90

WEBSTER'S AMERICAN MILITARY BIOGRAPHIES, edited by Robert McHenry. Over 1,000 figures who shaped 3 centuries of American military history. Detailed biographies of Nathan Hale, Douglas MacArthur, Mary Hallaren, others. Chronologies of engagements, more. Introduction. Addenda. 1,033 entries in alphabetical order. xi + 548pp. 6½ × 9¼. (Available in U.S. only)
24758-9 Pa. $13.95

LIFE IN ANCIENT EGYPT, Adolf Erman. Detailed older account, with much not in more recent books: domestic life, religion, magic, medicine, commerce, and whatever else needed for complete picture. Many illustrations. 597pp. 5⅜ × 8½.
22632-8 Pa. $8.95

HISTORIC COSTUME IN PICTURES, Braun & Schneider. Over 1,450 costumed figures shown, covering a wide variety of peoples: kings, emperors, nobles, priests, servants, soldiers, scholars, townsfolk, peasants, merchants, courtiers, cavaliers, and more. 256pp. 8⅜ × 11¼.
23150-X Pa. $9.95

THE NOTEBOOKS OF LEONARDO DA VINCI, edited by J. P. Richter. Extracts from manuscripts reveal great genius; on painting, sculpture, anatomy, sciences, geography, etc. Both Italian and English. 186 ms. pages reproduced, plus 500 additional drawings, including studies for *Last Supper, Sforza* monument, etc. 860pp. 7⅞ × 10¾. (Available in U.S. only) 22572-0, 22573-9 Pa., Two-vol. set $31.90

THE ART NOUVEAU STYLE BOOK OF ALPHONSE MUCHA: All 72 Plates from "Documents Decoratifs" in Original Color, Alphonse Mucha. Rare copyright-free design portfolio by high priest of Art Nouveau. Jewelry, wallpaper, stained glass, furniture, figure studies, plant and animal motifs, etc. Only complete one-volume edition. 80pp. 9⅜ × 12¼. 24044-4 Pa. $10.95

ANIMALS: 1,419 Copyright-Free Illustrations of Mammals, Birds, Fish, Insects, Etc., edited by Jim Harter. Clear wood engravings present, in extremely lifelike poses, over 1,000 species of animals. One of the most extensive pictorial source-books of its kind. Captions. Index. 284pp. 9 × 12. 23766-4 Pa. $10.95

OBELISTS FLY HIGH, C. Daly King. Masterpiece of American detective fiction, long out of print, involves murder on a 1935 transcontinental flight—"a very thrilling story"—NY Times. Unabridged and unaltered republication of the edition published by William Collins Sons & Co. Ltd., London, 1935. 288pp. 5⅜ × 8½. (Available in U.S. only) 25036-9 Pa. $5.95

VICTORIAN AND EDWARDIAN FASHION: A Photographic Survey, Alison Gernsheim. First fashion history completely illustrated by contemporary photographs. Full text plus 235 photos, 1840-1914, in which many celebrities appear. 240pp. 6½ × 9¼. 24205-6 Pa. $8.95

THE ART OF THE FRENCH ILLUSTRATED BOOK, 1700-1914, Gordon N. Ray. Over 630 superb book illustrations by Fragonard, Delacroix, Daumier, Doré, Grandville, Manet, Mucha, Steinlen, Toulouse-Lautrec and many others. Preface. Introduction. 633 halftones. Indices of artists, authors & titles, binders and provenances. Appendices. Bibliography. 608pp. 8⅜ × 11¼. 25086-5 Pa. $24.95

THE WONDERFUL WIZARD OF OZ, L. Frank Baum. Facsimile in full color of America's finest children's classic. 143 illustrations by W. W. Denslow. 267pp. 5⅜ × 8½. 20691-2 Pa. $7.95

FOLLOWING THE EQUATOR: A Journey Around the World, Mark Twain. Great writer's 1897 account of circumnavigating the globe by steamship. Ironic humor, keen observations, vivid and fascinating descriptions of exotic places. 197 illustrations. 720pp. 5⅜ × 8½. 26113-1 Pa. $15.95

THE FRIENDLY STARS, Martha Evans Martin & Donald Howard Menzel. Classic text marshalls the stars together in an engaging, nontechnical survey, presenting them as sources of beauty in night sky. 23 illustrations. Foreword. 2 star charts. Index. 147pp. 5⅜ × 8½. 21099-5 Pa. $3.95

FADS AND FALLACIES IN THE NAME OF SCIENCE, Martin Gardner. Fair, witty appraisal of cranks, quacks, and quackeries of science and pseudoscience: hollow earth, Velikovsky, orgone energy, Dianetics, flying saucers, Bridey Murphy, food and medical fads, etc. Revised, expanded In the Name of Science. "A very able and even-tempered presentation."—The New Yorker. 363pp. 5⅜ × 8. 20394-8 Pa. $6.95

ANCIENT EGYPT: Its Culture and History, J. E. Manchip White. From pre-dynastics through Ptolemies: society, history, political structure, religion, daily life, literature, cultural heritage. 48 plates. 217pp. 5⅜ × 8½. 22548-8 Pa. $5.95

SIR HARRY HOTSPUR OF HUMBLETHWAITE, Anthony Trollope. Incisive, unconventional psychological study of a conflict between a wealthy baronet, his idealistic daughter, and their scapegrace cousin. The 1870 novel in its first inexpensive edition in years. 250pp. 5⅜ × 8½. 24953-0 Pa. $6.95

LASERS AND HOLOGRAPHY, Winston E. Kock. Sound introduction to burgeoning field, expanded (1981) for second edition. Wave patterns, coherence, lasers, diffraction, zone plates, properties of holograms, recent advances. 84 illustrations. 160pp. 5⅜ × 8¼. (Except in United Kingdom) 24041-X Pa. $3.95

INTRODUCTION TO ARTIFICIAL INTELLIGENCE: Second, Enlarged Edition, Philip C. Jackson, Jr. Comprehensive survey of artificial intelligence—the study of how machines (computers) can be made to act intelligently. Includes introductory and advanced material. Extensive notes updating the main text. 132 black-and-white illustrations. 512pp. 5⅜ × 8½. 24864-X Pa. $10.95

HISTORY OF INDIAN AND INDONESIAN ART, Ananda K. Coomaraswamy. Over 400 illustrations illuminate classic study of Indian art from earliest Harappa finds to early 20th century. Provides philosophical, religious and social insights. 304pp. 6⅜ × 9⅜. 25005-9 Pa. $11.95

THE GOLEM, Gustav Meyrink. Most famous supernatural novel in modern European literature, set in Ghetto of Old Prague around 1890. Compelling story of mystical experiences, strange transformations, profound terror. 13 black-and-white illustrations. 224pp. 5⅜ × 8½. (Available in U.S. only) 25025-3 Pa. $6.95

PICTORIAL ENCYCLOPEDIA OF HISTORIC ARCHITECTURAL PLANS, DETAILS AND ELEMENTS: With 1,880 Line Drawings of Arches, Domes, Doorways, Facades, Gables, Windows, etc., John Theodore Haneman. Sourcebook of inspiration for architects, designers, others. Bibliography. Captions. 141pp. 9 × 12. 24605-1 Pa. $7.95

BENCHLEY LOST AND FOUND, Robert Benchley. Finest humor from early 30s, about pet peeves, child psychologists, post office and others. Mostly unavailable elsewhere. 73 illustrations by Peter Arno and others. 183pp. 5⅜ × 8½. 22410-4 Pa. $4.95

ERTÉ GRAPHICS, Erté. Collection of striking color graphics: *Seasons, Alphabet, Numerals, Aces* and *Precious Stones*. 50 plates, including 4 on covers. 48pp. 9⅜ × 12¼. 23580-7 Pa. $7.95

THE JOURNAL OF HENRY D. THOREAU, edited by Bradford Torrey, F. H. Allen. Complete reprinting of 14 volumes, 1837–61, over two million words; the sourcebooks for *Walden*, etc. Definitive. All original sketches, plus 75 photographs. 1,804pp. 8½ × 12¼. 20312-3, 20313-1 Cloth., Two-vol. set $130.00

CASTLES: Their Construction and History, Sidney Toy. Traces castle development from ancient roots. Nearly 200 photographs and drawings illustrate moats, keeps, baileys, many other features. Caernarvon, Dover Castles, Hadrian's Wall, Tower of London, dozens more. 256pp. 5⅜ × 8¼. 24898-4 Pa. $6.95

CATALOG OF DOVER BOOKS

AMERICAN CLIPPER SHIPS: 1833–1858, Octavius T. Howe & Frederick C. Matthews. Fully-illustrated, encyclopedic review of 352 clipper ships from the period of America's greatest maritime supremacy. Introduction. 109 halftones. 5 black-and-white line illustrations. Index. Total of 928pp. 5⅜ × 8½.
25115-2, 25116-0 Pa., Two-vol. set $17.90

TOWARDS A NEW ARCHITECTURE, Le Corbusier. Pioneering manifesto by great architect, near legendary founder of "International School." Technical and aesthetic theories, views on industry, economics, relation of form to function, "mass-production spirit," much more. Profusely illustrated. Unabridged translation of 13th French edition. Introduction by Frederick Etchells. 320pp. 6⅛ × 9¼. (Available in U.S. only) 25023-7 Pa. $8.95

THE BOOK OF KELLS, edited by Blanche Cirker. Inexpensive collection of 32 full-color, full-page plates from the greatest illuminated manuscript of the Middle Ages, painstakingly reproduced from rare facsimile edition. Publisher's Note. Captions. 32pp. 9⅜ × 12¼. 24345-1 Pa. $5.95

BEST SCIENCE FICTION STORIES OF H. G. WELLS, H. G. Wells. Full novel *The Invisible Man*, plus 17 short stories: "The Crystal Egg," "Aepyornis Island," "The Strange Orchid," etc. 303pp. 5⅜ × 8½. (Available in U.S. only) 21531-8 Pa. $6.95

AMERICAN SAILING SHIPS: Their Plans and History, Charles G. Davis. Photos, construction details of schooners, frigates, clippers, other sailcraft of 18th to early 20th centuries—plus entertaining discourse on design, rigging, nautical lore, much more. 137 black-and-white illustrations. 240pp. 6⅛ × 9¼.
24658-2 Pa. $6.95

ENTERTAINING MATHEMATICAL PUZZLES, Martin Gardner. Selection of author's favorite conundrums involving arithmetic, money, speed, etc., with lively commentary. Complete solutions. 112pp. 5⅜ × 8½. 25211-6 Pa. $3.50

THE WILL TO BELIEVE, HUMAN IMMORTALITY, William James. Two books bound together. Effect of irrational on logical, and arguments for human immortality. 402pp. 5⅜ × 8½. 20291-7 Pa. $8.95

THE HAUNTED MONASTERY and THE CHINESE MAZE MURDERS, Robert Van Gulik. 2 full novels by Van Gulik continue adventures of Judge Dee and his companions. An evil Taoist monastery, seemingly supernatural events; overgrown topiary maze that hides strange crimes. Set in 7th-century China. 27 illustrations. 328pp. 5⅜ × 8½. 23502-5 Pa. $6.95

CELEBRATED CASES OF JUDGE DEE (DEE GOONG AN), translated by Robert Van Gulik. Authentic 18th-century Chinese detective novel; Dee and associates solve three interlocked cases. Led to Van Gulik's own stories with same characters. Extensive introduction. 9 illustrations. 237pp. 5⅜ × 8½.
23337-5 Pa. $5.95

Prices subject to change without notice.

Available at your book dealer or write for free catalog to Dept. GI, Dover Publications, Inc., 31 East 2nd St., Mineola, N.Y. 11501. Dover publishes more than 175 books each year on science, elementary and advanced mathematics, biology, music, art, literary history, social sciences and other areas.